Chassis-Domain-Oriented Dynamic Control for Autonomous Vehicles

Over seven detail-rich chapters, this book comprehensively describes autonomous vehicle chassis modeling and control, chassis domain dynamic control, the estimation of essential dynamic states, research on motion planning, the development of chassis coordinated control, and related topics.

This book first summarizes vehicle dynamic modeling and control and provides the background and related topics for chassis domain dynamic control. It then presents the motivations of chassis domain control and introduces its conceptual framework. The book then focuses on the identification of tire–road interactions, which contain lateral, longitudinal, and vertical tire forces, before then discussing the estimation of essential dynamic states, which represent vehicle handling stability status, and the observation of road surface coefficient. The quantitative evaluation of vehicle chassis domain performance is then provided, with the rigorous definition and design of a comprehensive metric for assessing chassis dynamic performance. Next, the book instructs readers on the chassis-domain dynamic-aware motion planning for autonomous vehicles and the multi-objective multi-subsystem coordinated control. Finally, the authors present their conclusions and future recommendations for the advanced control of autonomous vehicles. The content and structure of this book will enable readers to address the high complexity and unpredictability of traffic conditions, along with the strong nonlinearity of vehicle dynamics during maneuvers, to facilitate the safe and coordinated operations of chassis subsystems. This will further the advancement of autonomous vehicles as the automobile industry transitions into the intelligent age.

This is a vital guide for readers from various expertise backgrounds. Advanced undergraduate and postgraduate students who study vehicle engineering will benefit from the descriptions of theoretical foundations and practical methodologies. Engineers and researchers will also benefit from the unique insights into modeling and control technologies for autonomous vehicles.

Chassis-Domain-Oriented Dynamic Control for Autonomous Vehicles

Estimation, Modeling, Motion Planning and Execution

Shuo Cheng, Peng Dong, Xiangyang Xu,
Shuhan Wang and Yanfang Liu

CRC Press
Taylor & Francis Group
Boca Raton London New York

CRC Press is an imprint of the
Taylor & Francis Group, an **informa** business

Designed cover image: Shutterstock

First published 2026
by CRC Press
2385 NW Executive Center Drive, Suite 320, Boca Raton FL 33431

and by CRC Press
4 Park Square, Milton Park, Abingdon, Oxon, OX14 4RN

CRC Press is an imprint of Taylor & Francis Group, LLC

ISBN: 978-1-032-95253-6 (hbk)
ISBN: 978-1-032-95254-3 (pbk)
ISBN: 978-1-003-58392-9 (ebk)

DOI: 10.1201/9781003583929

Typeset in Latin Modern font
by KnowledgeWorks Global Ltd.

Publisher's note: This book has been prepared from camera-ready copy provided by the authors.

Contents

Preface

The automobile industry is currently transitioning into the intelligent age, marked by disruptive advancements in automotive technology. Autonomous vehicles (AVs) represent the forefront of modern automobile science and technological innovation. Autonomous driving relies on intelligent chassis dynamic control because all vehicle motions are determined by the interactions between the chassis tires and the road. The high complexity and unpredictability of traffic conditions, along with the strong nonlinearity of vehicle dynamics during maneuvers, pose great challenges to the safe and coordinated operations of chassis subsystems. Vehicle dynamic nonlinearity makes it difficult to obtain vehicle dynamic states accurately. Complex road environments and uncertain traffic conditions can easily induce the instability of chassis dynamics. Additionally, with the rapid development of chassis electronics, the proliferation of chassis subsystems presents significant challenges in coordinating these subsystems to ensure dynamic stability. Addressing these opening issues is urgent. This book aims to systematically sort out these related topics and introduces the *chassis domain dynamic control* for AVs. It can provide a comprehensive understanding of the current developments and prospects of automotive chassis technologies, thereby facilitating the deployment of AVs.

This book includes eight chapters in total covering a comprehensive survey of vehicle chassis modeling and control, an introduction to chassis domain dynamic control, studies on the estimation of essential dynamic states, research on motion planning, the development of chassis coordinated control, etc. This book starts with an overview of vehicle dynamic modeling and control, which provides the background and related topics for chassis domain dynamic control. Chapter 2 presents the motivations of chassis domain control and introduces its conceptual framework. Chapter 3 focuses on the identification of tire–road interactions, which contain lateral, longitudinal, and vertical tire forces. In Chapter 4, This book will further discuss the estimation of essential dynamic states, which represent vehicle handling stability status, and the observation of road surface coefficient is also presented. Chapter 5 presents the quantitative evaluation of vehicle chassis domain performance, where the rigorous definition and design of a comprehensive metric for assessing chassis dynamic performance are provided. Furthermore, Chapter 5 focuses on the chassis-domain-dynamic-aware motion planning for AVs. Then, the multi-objective multi-subsystem coordinated control is introduced in Chapter 6. Finally, Chapter 7 discusses the conclusions and future recommendations for advanced control of AVs.

As a professional reference and monograph, this book details cutting-edge research into automotive chassis modeling and control technology and simultaneously provides a research basis and methods of investigating advanced technologies in automobile engineering. This book will benefit readers with a variety of expertise backgrounds. It can be used as a reference for graduate students in universities to systematically learn related knowledge bases of vehicle dynamic control, especially for AVs. This book will also provide developmental engineers, professional researchers, and university faculties of automotive engineering and control engineering with leading-edge and unique insight into modeling and control technologies for AVs. We hope this book will inspire innovative research to further facilitate the launch and advancement of AVs.

The contents of this book involve Dr. Cheng's research works conducted at Tsinghua University, China; the University of Tokyo, Japan; and Nanyang Technological University, Singapore. We are very grateful to Professor Liang Li, Professor Kimihiko Nakano, and Associate Professor Chen Lv, who offered Dr. Cheng professional and meticulous supervision. Thanks to Qingqing Su for her efforts in the process of manuscript submission. We are also thankful to CRC Press, and we would like to thank Editor Andrew Stow and Editorial Assistant Nabila Khadija Ansari for providing their support to complete this book.

Acknowledging the limitations of the authors, it is recognized that some imperfections may exist within this book. If possible, the subsequent revised version of the book will be provided in the coming years to ensure it remains up-to-date.

Beijing, China

Shuo Cheng
Peng Dong
Xiangyang Xu
Shuhan Wang
Yanfang Liu
May 2025

Author

Shuo Cheng is an associate professor at the School of Transportation Science and Engineering, Beihang University, Beijing, China. He received his B.S. degree from the Harbin Institute of Technology, Harbin, China, in 2016, and his Ph.D. degree in mechanical engineering from the School of Vehicle and Mobility, Tsinghua University, Beijing, China, in 2021. He was a postdoctoral fellow at the Institute of Industrial Science, the University of Tokyo, Tokyo, Japan, from 2021 to 2023. He was a research fellow at the School of Mechanical and Aerospace Engineering, Nanyang Technological University, Singapore, from 2023 to 2025. He serves as an associate editor for the SAE International Journal of Connected and Automated Vehicles and a young editorial board member for the Journal of Mechanical Engineering (Chinese). Dr. Cheng received the JSPS Postdoctoral Fellowship in 2021. His research interests include intelligent motion planning and control, automotive chassis system design, and dynamic control for autonomous vehicles.

Peng Dong is a professor at the School of Transportation Science and Engineering, Beihang University, Beijing, China. He received his B.S. degree in industrial engineering from Shandong University, Shandong, China, in 2008. He earned his M.S. degree in vehicle engineering from Beihang University, Beijing, China, in 2011. He received his Ph.D. degree in mechanical engineering from Ruhr-University Bochum, Germany, in 2015. His research interests include powertrain systems of electric vehicles, design and control of vehicle transmission, vehicle electric propulsion, hybrid energy management, and vehicle distributed drive technology.

Xiangyang Xu received the B.S. degree and the M.S. degree in vehicle engineering, Beijing Institute of Technology, Beijing, China, in 1987 and 1990, respectively, and the Ph.D. degree in mechatronic engineering, Harbin Institute of Technology, Harbin, China, in 1999. From 1998 to 1999, he was a Visiting Scholar at Daimler AG, Stuttgart, Germany. From 1990 to 2002, he is a staff of the College of Automotive Engineering, Harbin Institute of Technology. He has been a professor at the School of Transportation Science and Technology, Beihang University, since 2002. Prof. Xu is a member of the Advisory Board, Innovation Automotive Transmissions, Hybrid & Electrical Drives, International CTI Symposium and Exhibition (Berlin, Shanghai). He serves as a Fellow of SAE-China, an Executive member of SAE-China, a Deputy Secretary-General for SAE-China of Gear Technology Branch, a Deputy Secretary-General for China Gear Manufacturer Association, the Director of Beijing Key Lab for High Efficiency and System Control of New Energy Resource Vehicle, and an Executive Deputy Director of National

Engineering Research Center for Passenger Car Automatic Transmission. He was awarded a Best Paper Award by the 2019 SAE International Intelligent and Connected Vehicles Symposium. His research interests include transmission technology, hybrid, and EV drive.

Shuhan Wang received a B.S. degree in mechanical design and manufacturing and automation from the Harbin Institute of Technology, Harbin, China, in 2002, and a Ph.D. degree in vehicle engineering from Beihang University, Beijing, China, in 2009. From 2007 to 2008, he was a joint Ph.D. student with the Chemnitz University of Technology, Germany. He is currently a professor at the School of Transportation Science and Engineering, Beihang University, and the Deputy Director of the National Passenger Vehicle Automatic Transmission Engineering Technology Research Center. In 2016, he won the First Prize of the National Science and Technology Progress Award (ranked third).

Yanfang Liu is a professor at the School of Transportation Science and Engineering, Beihang University, Beijing, China. She received her B.S. degree in mechanical design and manufacture from the Qilu University of Technology, Jinan, China, in 2000, and her Ph.D. degree in mechanical engineering from Beihang University, Beijing, China, in 2005. She serves as the Deputy Director of the Expert Committee of the Simulation Technology Industry Branch of the China Industrial Cooperation Association. Her research interests include the collaborative design and optimization of automotive transmission systems, vehicle multi-domain system dynamics and its intelligent control, etc.

1 Introduction

This chapter first sheds light on the significance and motivation of chassis domain dynamic control for autonomous driving by analyzing the background and challenges in vehicle dynamics, especially in the context of self-driving. Then, the fundamental modeling of vehicle dynamics is briefly introduced. Motion planning and dynamic control will also be discussed. Finally, the overall structure of this book will be briefly discussed.

1.1 SIGNIFICANCE AND MOTIVATION

The automobile industry, as one of the core industries of manufacturing, is a crucial pillar of the national economy. It has entered the era of intelligent development, bringing about disruptive changes in automotive technology. With the rapid development of sensing, electronic information, electronic control, and artificial intelligence, innovation centered on autonomous driving has become the focal point of the new industrial revolution and technological competition. Substantial efforts in developing key technologies for autonomous driving from automotive manufacturers and research institutions have been reported [87, 75, 8]. As a critical assembly, vehicle chassis is at the core of the implementation of autonomous driving. The stability of chassis dynamics is crucial for the safe and stable operation of a car [135].

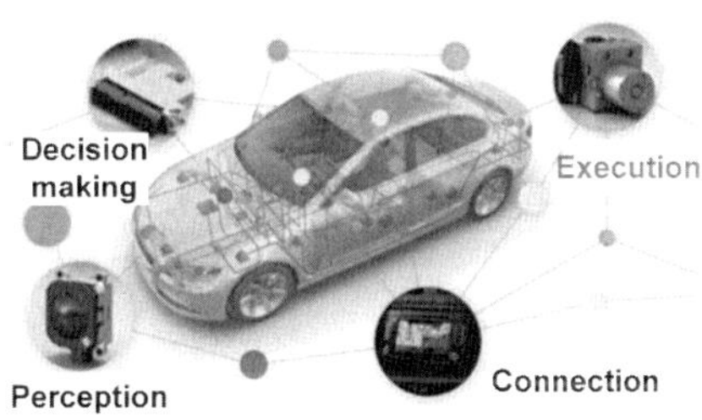

Figure 1.1 Major components of AVs.

As illustrated in Fig. 1.1, autonomous vehicles (AVs) use onboard sensors together with communication and network technologies to perceive surrounding traffic and environments such that this perception can be processed and analyzed to make driving decisions and conduct motion planning. Then, various chassis components ultimately execute those upper commands. To clarify the differences between various levels of autonomous driving technologies, the SAE and NHTSA have classified and defined the levels of vehicle autonomy [117]. Autonomous driving functions below Level 3, mainly referred

DOI: 10.1201/9781003583929-1

to as ADAS, can significantly enhance vehicle active safety and serve as the foundation for achieving fully self-driving capability. In recent years, major car manufacturers have increased their investments to integrate ADAS into various vehicle models for mass production. Common ADAS functions can be categorized as follows: AEB, LKA, ALC, and ACC [101]. The implementation of these ADAS functions relies on the coordinated execution of chassis components involving braking, traction, and steering systems. Notably, all vehicle motions are determined ultimately by the tire forces of chassis wheels. Therefore, vehicle chassis is crucial for the control and implementation of decision-making and planning in autonomous driving and ADAS.

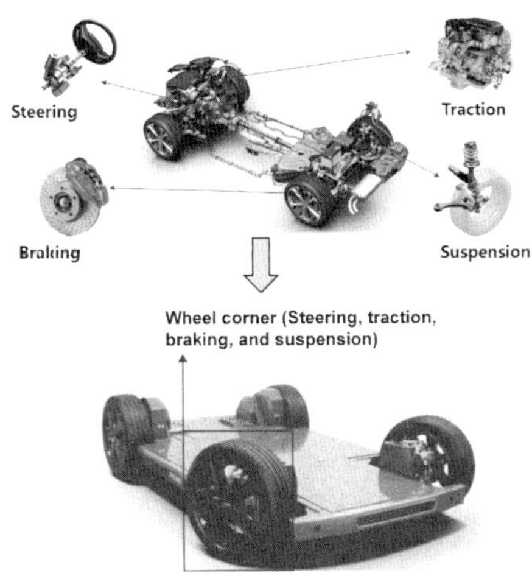

Figure 1.2 Demonstration of automotive chassis systems.

With the rapid advancement of sensor technology, electronic technology, and advanced control technology, automotive chassis systems are transitioning from purely mechanical systems to integrated electromechanical control systems, as illustrated in Fig. 1.2. This transformation provides the foundational implementation platform for autonomous driving functions and is essential to ensure the safe and stable operations of AVs. With regard to chassis electronic control, major manufacturers and research institutions worldwide continuously develop various active safety controllers in vehicle chassis to improve dynamic stability and enhance vehicle driving safety. The functionality of ABS is to maximize road adhesion during hard braking, thereby enhancing vehicle driving stability during deceleration [136, 137]. TCS aims to improve vehicle traction and stability performance through active adjustments of the EMS and HCU such that the wheel slip rates of traction wheels are within the region of maximum adhesion [33, 138]. DYC first assesses vehicle handling stability based on the driver's maneuvers and actual vehicle response to identify over-steering or under-steering status. Then, the required active braking pressures for each wheel can be calculated through the active yaw moment control strategy, which is executed by the HCU. Thus, DYC can correct the vehicle's unstable dynamic state, thereby enhancing vehicle dynamic stability during steering. Since Bosch first equipped Mercedes-Benz cars with ABS in 1978, decades of development have led to the successive introduction of TCS and DYC functions [128]. Then, all these systems have been integrated into ESP. This integrated control system significantly improves the coordination of ABS, TCS, and DYC functionalities. According to statistics from the previous report [43], vehicles equipped with ESP have seen a 41% reduction in the number of single-vehicle accidents and a 56% reduction in fatal single-vehicle accidents. Therefore, the emergence of ESC is of great significance for vehicle active safety and the integration of chassis electronic control systems.

Furthermore, functions in ADAS mentioned above are designed separately to enhance specific vehicle performance aspects such as emergency obstacle avoidance, safe car following distance, lane keeping, and safe lane changing. However, the operational functioning of these subsystems depends on maintaining chassis dynamic stability. For instance, ACC aims to adjust the ego vehicle's longitudinal acceleration to ensure safe car-following distances. However, the unpredictability of the leading vehicle's motion can push the ego vehicle into the region of dynamic nonlinearity, potentially causing instability such as the vehicle's large sideslip or sharp spin [27]. Similarly, AEB engages high-intensity braking during emergency maneuvers, but the road surface uncertainties, for example, split friction conditions, might lead to large sideslip or fishtailing [24]. LKA and ALC subsystems control the steering angle to maintain lane position or execute lane-change maneuvers. During these maneuvers, factors such as large steering angles, load transfer, and low-adhesion road conditions can push vehicle dynamic evolution into nonlinear regions, which might make the ego car reach its tire adhesion limits and cause dynamic instability. Although ESC can intervene to enhance chassis stability when encountering these dangerous situations, excessive braking intervention can degrade the performance of ADAS functions and negatively impact ride comfort.

As illustrated in Fig. 1.3, over the century-long development of vehicle dynamics, the automotive chassis has evolved through phases of the innovations of chassis component design, matching and calibration of dynamic performance, and breakthroughs in electromechanical dynamic control. These advancements have progressively addressed key dynamic issues arising from the increase in vehicle speeds such that vehicle driving performance has been further enhanced. Currently, an automobile is a complex electromechanical system comprising numerous components and subsystems, and each subsystem is mainly designed to optimize a specific performance metric. However, the current isolated design methodology can lead to inaccuracies in dynamic modeling since all these subsystems only utilize specific reduced-dimension models that are related to their control objectives. During vehicle operations, most

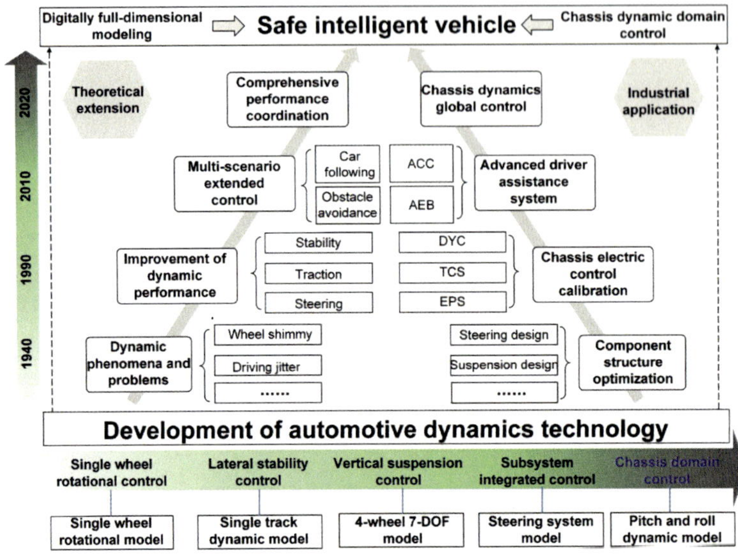

Figure 1.3 Development history of automotive chassis engineering.

subsystems interact and constrain each other, posing challenges in control coordination. Thereby, enhancing overall vehicle dynamic performance depends on the seamless cooperation of all these subsystems. With the trend toward high-speed and efficient vehicle operation, automotive dynamics face significant challenges in adapting to the disruptive changes brought by autonomous driving. The increasing integration of electromechanical components in vehicle chassis necessitates a shift from single-objective control to multi-system and multi-objective integrated control, which motivates my research ideas on chassis domain control for AVs. Additionally, the digitalization and intelligentization of automotive dynamics are essential for the launch of AVs. Unlike traditional vehicles, AVs lack human drivers' real-time perception and control of dynamic states, leaving no immediate intervention for any potential risks including dynamic instability. Additionally, the absence of human drivers in self-driving cars presents significant challenges to current traffic laws, as there is no individual to hold accountable for car accidents. Therefore, it is crucial to explore mechanisms for digitalizing, accurately computing, and predicting dynamic evolutions of AVs, as well as coordinating control systems for the improvement of AVs' safety. Thus, chassis dynamic domain control is a critical technology that must be developed to ensure the successful deployment and application of AVs.

Chassis domain dynamic control leverages shared sensor signals from various chassis components and then operates the integrated observation of vehicle motion and dynamic states to achieve an accurate full-dimension vehicle dynamic model rather than dedicated reduced-dimension models designed by current chassis subsystems. By establishing the full-dimension model and unifying its solution to depict vehicle dynamic evolutions, this approach digitalizes chassis dynamics, which can potentially overcome common challenges in the AV industry. This domain control system can coordinate and optimize multiple objectives including subsystem control functions and dynamic stability, across all chassis actuators. Thus, chassis components can be transformed into actuators that only execute specific actions, which eliminates the requirements for tedious calibration and matching for each subsystem's ECU and control strategy. To achieve this, it is crucial to accurately obtain chassis global dynamic states and parameters from limited sensor signals in real-time. These key parameters include tire forces on each wheel in all directions, which are essential for vehicle dynamic modeling, tire dynamic parameters such as tire sideslip angle and road friction coefficient, which provide critical information for subsystems to adapt to different road conditions, and vehicle sideslip angle, a fundamental feedback control variable for dynamic stability control. Based on the global information, an accurate full-dimension vehicle dynamic model can be established, providing a foundational model for chassis domain control. Subsequently, the control objectives of various subsystems and the performance of dynamic stability are required to be analyzed. We must design elaborately the architecture of chassis dynamic domain control to comprehensively optimize various chassis performance aspects. Furthermore, investigations on coordination control algorithms are essential to achieve integrated and coordinated control of multiple advanced driving functions and dynamic stability. This book will detail these key aspects in the following chapters, and the following sections of this chapter will provide brief introductions about these issues.

1.2 VEHICLE DYNAMIC MODELING

To achieve precise control of vehicle dynamic evolutions, the first and fundamental step is to establish vehicle dynamic models, which are required to derive specific control laws. We will not recapitulate all aspects of vehicle modeling in this book and will deliberately concentrate on related aspects of dynamic control. For vehicle dynamic control research, prevailing dynamic models are the single-track bicycle model depicting vehicle lateral and yaw motions, the 3 DOF dynamic model describing longitudinal, lateral, and yaw motions, the four-wheel 7 DOF dynamic model, the quarter vehicle model with one dimension representing vehicle vertical vibrations, etc. In addition, vehicle tires are the links between a car and the road surface, which not only support vehicle loads but also generate forces controlling vehicle motions. Thus, we must consider the modeling of tire dynamics when studying vehicle dynamic control. Existing tire models can be classified into three types: pure physical models, empirical dynamic models, and semi-empirical models [105]. For vehicle dynamic control, the prevailing tire models contain the linear tire model, uni-tire model [53], Dugoff tire model [40], magic formula tire model [106], etc. (Refer to the literature [145, 12, 51] for details.) It is necessary to accurately estimate crucial states and parameters, which provides a foundation for the above-mentioned dynamic modeling. For instance, tire forces are critical parameters for vehicle modeling, the road-tire interaction information is essential for ensuring the road adaptability of various chassis subsystems, and vehicle sideslip angle is a crucial feedback control variable for dynamic stability control. This section will present some brief introductions about vehicle parameters and state estimation.

1.2.1 TIRE FORCES ESTIMATION

The tire print, also called contact patch, contact region, or tire footprint, denotes the contact area between vehicle tires and the road. Vehicle motions in all directions are primarily governed by tire forces transmitted through tire prints of vehicle wheels; thereby, tire forces play a dominant influence on vehicle dynamics and are critical parameters for vehicle dynamics control. Additionally, during extreme maneuvers, vehicle tire forces exhibit nonlinear characteristics relative to wheel longitudinal slip rates and slip angles. Numerous uncontrollable external factors, such as vehicle load, tire pressure and wear, and tire–road interface profile, significantly impact tire forces [114]. Accurately obtaining tire forces remains an open challenge.

Currently, there are three primary methods for obtaining tire forces: direct measurement, calculation via physical or empirical tire models, and estimation by designing observation algorithms [116, 126, 81].

★ Direct measurement of tire forces requires complex and specialized tire force sensors, such as the six-component wheel force transducer produced by Kistler Group (see Fig. 1.4). However, these sensors are prohibitively expensive and are not suitable for implementation on ordinary passenger cars.
★ Various tire models, such as the Dugoff model and the Pacejka tire model, have been extensively studied. We can use these tire models to calculate tire forces. However, these models necessitate extensive experimental calibration for different tire specifications and pressures. Additionally, the calibrated parameters are subject to change as the tires age.
★ Consequently, it is essential to develop tire force observation algorithms that do not depend on prior knowledge of tire–road characteristics or costly tire force sensors.

Substantial investigations on tire force estimations have been reported, and readers can refer to the review literature [76] to find details. Generally, the procedure of tire force estimation is that we design an observer based on a specific filter algorithm and utilize some signals such as a_x, a_y, and φ, which are measured by vehicle onboard sensors, to achieve estimated values of tire

Figure 1.4 Demonstration of the six-component wheel force transducer. The picture can be found on the website: https://www.kistler.com/US/en/tire-testing/C00000072.

forces. Common algorithms, for example, SMO, EKF, UKF, and RLS, have been utilized to design tire force estimators [51, 76].

1.2.2 VEHICLE SIDESLIP ANGLE ESTIMATION

Vehicle sideslip angle is defined as the angle between the vector of vehicle velocity at the vehicle mass center and the x-axis of the vehicle coordinate system [16]. The magnitude of the vehicle sideslip angle indicates the dynamic stability level: if the vehicle sideslip angle is excessively large, beyond a certain threshold, vehicle tires will reach their adhesion saturation and lateral tire forces may not effectively control vehicle motions, indicating vehicle dynamic instability [30]. We argue that vehicle tire slip characteristics are fundamental to investigations on vehicle handling and stability. Concretely, the slip characteristics of the left and right wheels determine the slip characteristics of the corresponding vehicle axle. Furthermore, the combined slip characteristics of the front and rear axles determine the steering characteristics of the vehicle. Vehicle chassis control systems must actively intervene through braking, steering, and driving to maintain moderate under-steering, thereby ensuring dynamic stability.

Vehicle sideslip angle can be directly measured using devices such as the differential GPS. Its observation methods generally can be divided into two categories: integral observers based on sensor measurement signals and observation algorithms based on vehicle dynamic models [23]. Regarding tire sideslip angles, they can be estimated through direct visual measurement methods, observations based on steering system responses, and estimations based on the vehicle's dynamic response.

★ Vehicle sideslip angle can be directly measured using GPS equipment: dual-antenna differential GPS can measure both the direction and magnitude of the velocity vector at the vehicle mass center such that vehicle sideslip angle can be directly calculated. However, direct measurement requires additional GPS equipment to be mounted on the vehicle, and the high cost of these devices makes it impractical for mass production.

★ The basic principle of the integral observation algorithm based on measured signals by onboard sensors is as follows: We can integrate the signal of vehicle lateral acceleration measured by the onboard gyroscope to obtain vehicle lateral velocity. Then, the vehicle sideslip angle can be directly calculated by the formula $\beta = v_y/v_x$. The integral observation method does not require an accurate vehicle dynamic model and is highly robust to environmental changes. However, noises in the measured signals accumulate in the process of integration, which leads to cumulative observation errors.

★ Observation algorithms based on vehicle dynamic models derive the theoretical value of vehicle sideslip angle directly from the dynamic model as an estimate. These model-based observers can achieve high accuracy when a vehicle is operating in the linear region and are not affected by sensor errors. However, vehicle parameters in the dynamic model are influenced by factors including vehicle load, load distribution, tire characteristics, and tire wear, which introduces errors into the dependent dynamic model. Additionally, the vehicle model's error can increase significantly when the car enters its dynamic nonlinear region.

1.2.3 ROAD FRICTION COEFFICIENT ESTIMATION

Accurate identification of road surface friction coefficient enables various active safety control systems and advanced driving functions to dynamically adjust their control strategies in real-time. This enhances the effectiveness of active interventions, thereby improving vehicle dynamic stability and safety. The estimation of road surface friction is a crucial issue in chassis dynamic control; hence this issue has attracted extensive studies conducted by researchers worldwide in recent years [55, 91]. Significant progress in road surface friction identification through sensor technology, model parameter estimation, and various data analysis methods has been reported (Readers can refer to the review literature [1, 111]). The methods for observing road friction coefficients can be classified into direct detection methods based on sensors and indirect estimation methods based on the patterns of vehicle dynamic evolution [119].

★ Research on sensor-based direct detection methods, such as optical, acoustic, microwave, and image sensors, began as early as the 1990s. For instance, some researchers employed optical sensors to measure the absorption and scattering of light on the road surface, identifying the presence of water, ice, and other substances that diminish road friction [50]. This approach can detect changes in road surface friction.

★ Indirect observation methods based on vehicle dynamic models do not require the installation of various sensors on the vehicle. Instead, they can achieve the estimated road surface friction coefficient via the synthesis of vehicle dynamic states. Many researchers have contributed significantly to this technical route. We can use the tire slip ratio versus the friction coefficient curve to perform this estimation procedure. To be specific, the first step is to calculate the wheel slip ratio by using the wheel speed and vehicle speed information, and then we can estimate the road friction coefficient based on the slope of the slip ratio versus the friction coefficient curve in the linear region [54, 78].

1.3 MOTION PLANNING FOR AUTONOMOUS DRIVING

Motion planning, which originated in the 1960s, was initially applied in the robotics field to generate collision-free paths to a target position in open environments (significant advancements can be found in the surveys [96, 52]). For AVs, motion planning is required to perform a real-time transition from one feasible state to the next, considering vehicle dynamics and kinematic constraints, such that a safe, stable, and comfortable path can be generated [46]. Unlike traditional robotic systems, AVs face the challenge of navigating through uncertain traffic environments in intricate and unpredictable road conditions. Moreover, the planned trajectory must satisfy constraints related to vehicle dynamics and traffic regulations. Various investigations tailored to the characteristics of AVs and related traffic scenarios have been conducted [46].

- ★ The fundamental principle of graph search-based trajectory planning methods is to search for an optimal trajectory within a given state space according to specified rules or objective functions. Graphs can be weighted or directed using sampling points, grids, or trajectory nodes [64]. Common graph search-based trajectory planning algorithms include the Dijkstra algorithm, the A-star algorithm, and their derivatives [58, 125]. These methods are suitable for both structured and unstructured road scenarios. However, the stochastic nature of the search process introduces variability, and the size of the search space and the complexity of the environment significantly impact the efficiency of graph search.
- ★ Unlike graph search-based algorithms performing continuous expansion of new feasible nodes in space, curve interpolation methods construct a trajectory that satisfies the requirements of comfort and feasibility through interpolation based on a series of known reference points [77]. Common interpolation curves used in trajectory planning include straight-line-arc connections, Euler spirals, polynomial curves, and Bezier curves [110, 113].
- ★ The fundamental concept of space sampling-based trajectory planning algorithms is to sample and select trajectories that meet the specific requirements from a decomposed evolution space. Compared to

graph search-based methods, space sampling algorithms have probabilistic completeness such that they are suitable for solving trajectory planning problems in high-dimensional evolution spaces [41]. There are three main types of space decomposition: random sampling decomposition, connected cells decomposition, and lattice [32].

★ Numerical optimization-based motion planning methods usually construct a numerical optimization problem based on environmental constraints, vehicle status, and performance indicators of trajectory planning, and then perform iterative optimization to minimize or maximize a cost function with multiple constraints. For example, MPC-based methods have been reported to perform trajectory planning for AVs [44].

1.4 VEHICLE DYNAMIC CONTROL

The concept of dynamic stability emerged in the late 20th century and has advanced significantly through substantial efforts of automotive companies and research institutions worldwide [156, 121, 15]. This progression has moved from theoretical analysis and simulation validation to real-world vehicle implementation and industrial production. Since the 1980s, the rapid development of sensor technology, electronics, and automatic control technologies has facilitated the emergence of various vehicle electronic control systems [67]. These systems, such as ESC, AEB, ACC, LKA, and active suspension systems, can be categorized based on the vehicle's motion direction into longitudinal control (traction and braking control), lateral control (steering and yaw moment control), and vertical control (suspension control).

In the context of autonomous driving, the vehicle chassis, as the foundational execution platform, plays a pivotal role in ensuring safe and stable control. Thus, chassis dynamic control issues face unprecedented new challenges and have gained new significance. Additionally, most automotive electronic control systems are designed to address specific functions, acting through the tire prints. However, a vehicle's various directional movements are not independent, and all subsystems influence and constrain each other during maneuvers. Therefore, chassis-integrated and coordinated control, especially the coordinated control of dynamic stability, has been a complex and prominent topic attracting increasing attention from both academia and the automotive industry [133]. Currently, integrated control architectures can be divided into three types: the architecture of decentralized control in parallel, the centralized control architecture, and the hierarchical control architecture [148].

★ In the architecture of decentralized control in parallel, each chassis subsystem operates independently to optimize its specific control objectives. Communication between subsystems, if necessary, occurs through the vehicle CAN such that a certain degree of integrated control can be achieved. However, this integration can only be completed by the Original Equipment Manufacturers.

★ The centralized control architecture consolidates the control inputs of all chassis subsystems through a single global controller. This system is structured from the top down, considering the active control characteristics and objectives of the entire chassis. This architecture enables a high degree of integration and thorough contemplation of the vehicle chassis. However, it requires considerable computational resources and storage capacity for control system design.

★ The hierarchical control architecture has emerged as an effective strategy for managing the integration of multiple chassis subsystems. It can comprehensively consider various aspects of chassis performance and enhance overall vehicle performance so that the integrated chassis control system becomes more flexible and reliable. This architecture combines the advantages of both the architecture of decentralized control in parallel and the centralized control architecture. It has clear and independent layers that facilitate system development and functional expansion, thereby it has been widely adopted.

1.5 ORGANIZATION OF THIS BOOK

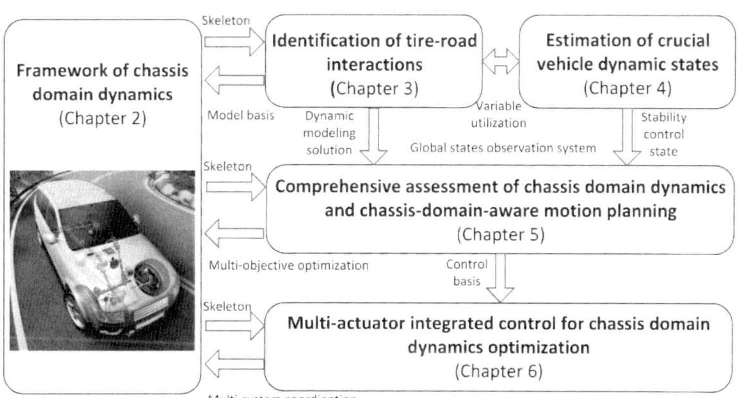

Figure 1.5 Overall structure of this book.

Fig. 1.5 illustrates the whole structure of this book. We organize the contents of this book into eight chapters, containing the whole aspects of chassis dynamics. First, Chapter 2 presents the concept and framework of chassis domain dynamic control, which provides an outline of the following chapters and the skeleton of our studies. Second, we summarize our research works on identifications of tire–road interactions in Chapter 3, which can be used as the model basis. The following chapter continues to further introduce related aspects of vehicle dynamic modeling and summarizes our work on the estimation of vehicle dynamic states. Then, the assessment of chassis domain dynamics and chassis-domain-aware motion planning is presented in Chapter 5, which can be used as the control basis. Finally, Chapter 6 details the multi-actuator integrated control method.

2 Concept and Framework of Chassis Domain Dynamics

This chapter focuses on the conceptual design and overall framework of chassis domain dynamic control for AVs. In the context of the rapid development of autonomous driving, this chapter first introduces the research background and analyzes essential chassis-relevant issues of the deployment of autonomous driving. This chapter presents the conceptual architecture and overall framework of the chassis domain dynamic control, which aims to provide a new solution to realize a more efficient, reliable, and flexible chassis platform supporting the launch of AVs. Analysis of vehicle dynamics is also introduced.

2.1 BACKGROUND

The vehicle chassis is a sophisticated electromechanical system assembled from numerous electronically controlled components such as the steering subsystem, braking subsystem, driving subsystem, and suspension subsystem. Each key component in a newly developed vehicle model incurs significant expenses for bespoke calibration and development. From the perspective of the essential mechanism, current electronic control components are designed in isolation based on different reduced-dimension models before being assembled in the chassis, as illustrated in Fig. 2.1. Each component consists of its own required sensors, calibrated dynamic models, control strategy algorithms, and actuators. This design procedure leads to several issues:

1. The suppliers of chassis components are usually different and fragmented. Each component requires individual development and specific calibration to operate normally, which results in substantial costs of development and calibration.
2. The proprietary communication protocols of different suppliers result in information silos between sensors and thus reduce the accuracy of reduced-dimension dynamic models used in those components.
3. The disparate design of each chassis subsystem, each with distinct optimization objectives might cause control mismatches, leading to inherent instability in chassis operation.

Interesting investigations on chassis coordinated control have been conducted, which aim to employ coordinated controllers or integrated control methods to comprehensively manage various subsystems such that the overall performance of vehicles can be improved [93]. Our previous work developed a robust H_∞ controller to coordinate the vehicle steering actuator and braking actuator via DYC and AFS such that the performance of vehicle dynamic

DOI: 10.1201/9781003583929-2

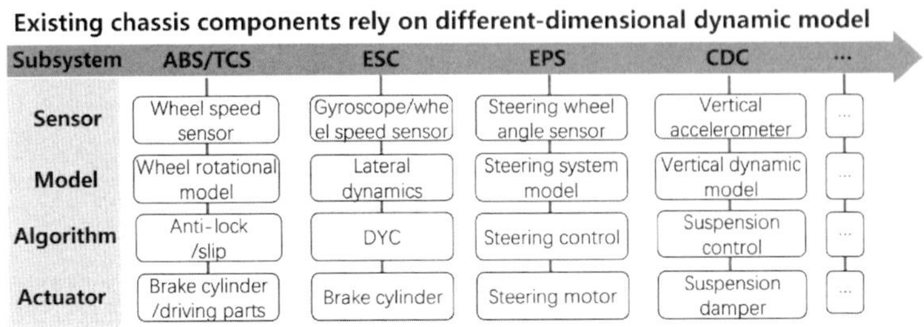

Figure 2.1 Existing chassis components rely on specific reduced-dimension dynamic models.

stability can be improved [26]. The developed controller is robust against the uncertainties of vehicle parameters, for example, the vehicle mass and vehicle longitudinal speed. Literature [152] designed a three-layer hierarchical integrated controller, which aims to coordinate the interactions among suspensions, AFS, and DYC. This study attempted to improve vehicle performance of both lateral and vertical dynamics. From the perspective of control architecture, we can summarize three types of chassis-coordinated control architectures, which have been briefly introduced in Section 1.4. In the following, details of each architecture, especially advantages and disadvantages, will be discussed.

2.1.1 ARCHITECTURE OF DECENTRALIZED CONTROL IN PARALLEL

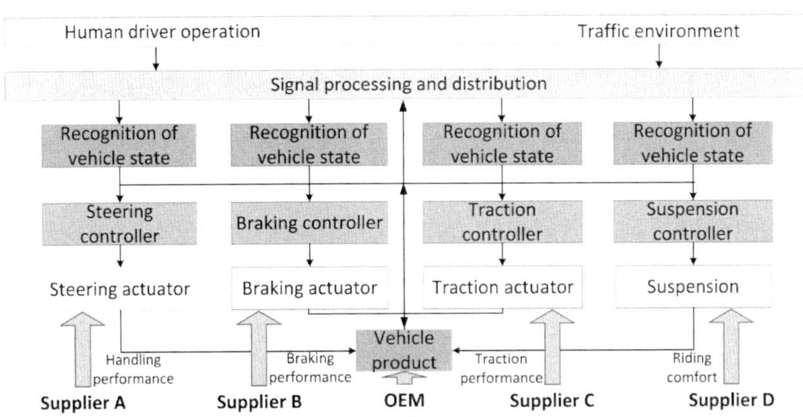

Figure 2.2 Demonstration of the architecture of decentralized control in parallel.

Early research on chassis-integrated control often focused on this control architecture, as shown in Fig. 2.2. Significant contributions from companies, for instance, Mitsubishi and Toyota, have been reported [127, 95], and those investigations were based on this architecture. This architecture focuses on the integration and measured signal sharing of onboard sensors across various chassis subsystems. Different suppliers only focus on the design and development of specific chassis components and provide individual components for OEM. The OEM performs vehicle assembly tasks and seeks to achieve the integration of several functions or communications. With the development of autonomous driving technologies, the proliferation of chassis electronic control subsystems has resulted in mutual interference among those subsystems, which limits the overall riding performance. There is no global control objective in this architecture, and it also suffers from poor information sharing between subsystems. Hence, this methodology does not apply to autonomous driving.

2.1.2 CENTRALIZED CONTROL ARCHITECTURE

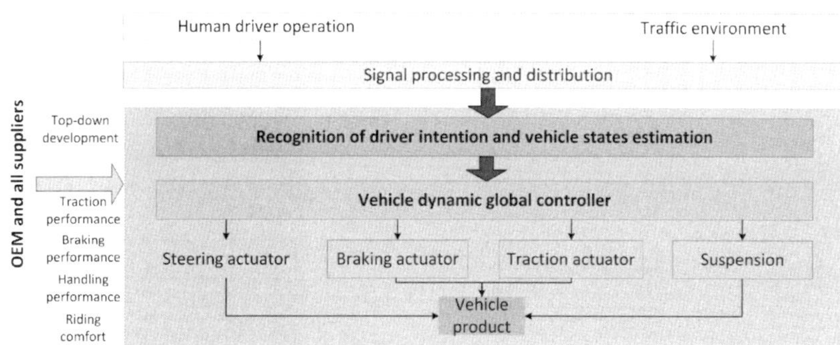

Figure 2.3 Demonstration of the centralized control architecture.

The top-down-designed centralized control architecture is illustrated in Fig. 2.3. In this architecture, the OEM and all suppliers are jointly responsible for achieving overall performance aspects of a car product, including traction performance, braking performance, handling performance, and riding comfort. All signals from onboard sensors and control strategies of various functions are integrated into one single global controller such that full integration of the entire chassis is achieved. Literature [99] employed this centralized control architecture and proposed an integrated control system. ARS and DYC were coordinated by this integrated control system to further improve vehicle handling performance. J. Andreasson et al. proposed to achieve global chassis control by means of model inversion-based feedforward with allocation directly on the actuator commands [3]. This work attempts to perform the desired vehicle motions while minimizing the utilization of the tire grip potential, which is achieved by sampling the constrained least-squares optimization of the constructed linearized problem. However, this high degree of integration poses huge challenges to computing hardware and storage devices. In addition, this architecture lacks flexibility and scalability. We must redesign the global controller if there is a new control subsystem or functionality that is required to be integrated into the existing chassis.

2.1.3 HIERARCHICAL CONTROL ARCHITECTURE

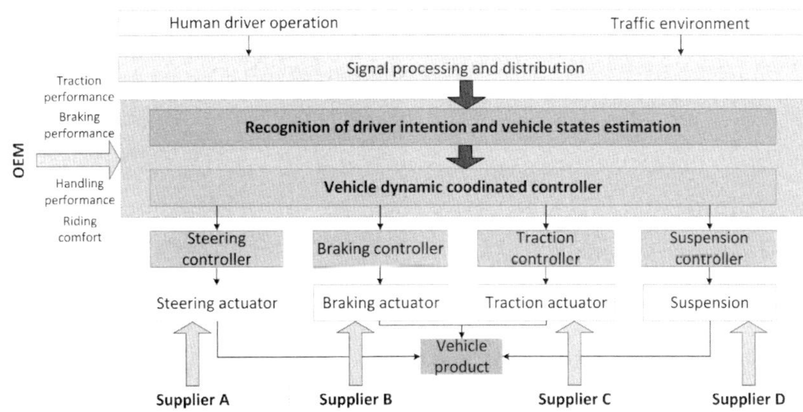

Figure 2.4 Demonstration of the hierarchical control architecture.

As illustrated in Fig. 2.4, this architecture transforms complex multi-objective control problems into relatively simpler control optimization problems. Human driver intention is identified based on inputs from driver operations. The onboard sensors then measure key signals of vehicle dynamic states, and specific state estimators further observe some unmeasurable variables. The chassis dynamic coordinated controller sends corresponding control commands to each chassis subsystem controller, and its actuator performs the desired action. In 2001, the Ford Motor Company proposed a similar architecture for vehicle chassis dynamic integrated control [14]. The literature [151] also designed an overall architecture for nonlinear integrated control of vehicle chassis based on hierarchical control, which includes an integrated control layer, a control allocation layer, and an actuator layer. In this architecture, the OEM takes charge of vehicle assembly and develops a vehicle dynamic coordinated controller based on all performance aspects. In contrast to the centralized control architecture, this hierarchical architecture requires the complete supply chain of all chassis actuators and their controllers from specific suppliers. Thus, the advantages of both the architecture of decentralized control in parallel and the centralized control architecture are present in this approach.

2.2 CONCEPTUAL ARCHITECTURE

This book introduces the concept of chassis dynamic domain control for AVs and proposes the overall technical solution for chassis dynamic domain control, as illustrated in Fig. 2.5. The core idea of this technical solution is as follows:

★ The processing procedure of all signals measured by chassis sensors is transferred to the chassis domain control system. This chassis dynamic domain solution uniformly receives signals from onboard sensors of all chassis components and performs signal fusion.

★ Crucial dynamic states and parameters are estimated based on the results of signal fusion. Then, it conducts precise iterative solving and predicting of the vehicle dynamic model through global dynamic state observation and full-dimension modeling. Dynamic features of vehicle longitudinal, lateral, vertical, yaw, rolling, and pitching motions are distributed to corresponding control functions by the proposed chassis domain system.

★ Based on dynamic modeling, the chassis domain system also accurately quantifies the boundary conditions for the safe and stable operation of AVs and quantitatively assesses the comprehensive performance of vehicle chassis. Simultaneously, it feeds back information, such as predictions of vehicle dynamic states and quantified stability conditions, to the decision-making module.

★ It performs chassis-domain-aware motion planning based on the decision-making instructions and dynamic status. Thus, an optimal and dynamically feasible trajectory, which satisfies geometrical smoothness, chassis global dynamics indices, and feasible safety, can be generated and transformed.

★ Then, our domain control system performs multi-objective and multi-system comprehensive optimization and coordinates the control of various chassis actuators.

★ All control commands are executed by the respective chassis actuators, including the braking actuator, traction actuator, steering actuator, and suspension actuator.

2.3 OVERALL FRAMEWORK

Based on the core idea of this technical solution as discussed in Section 2.2, we elaborately design the overall framework and logical flow for realizing the chassis dynamic domain control system and develop corresponding implementations for all components of the chassis domain system. As demonstrated in Fig. 2.6, we summarize all components of the chassis domain system into four submodules, which are covered in the following chapters of this book. In addition, we have also constructed two platforms to conduct the necessary experiments such that the effectiveness and performance of our methods can be validated and iteratively optimized.

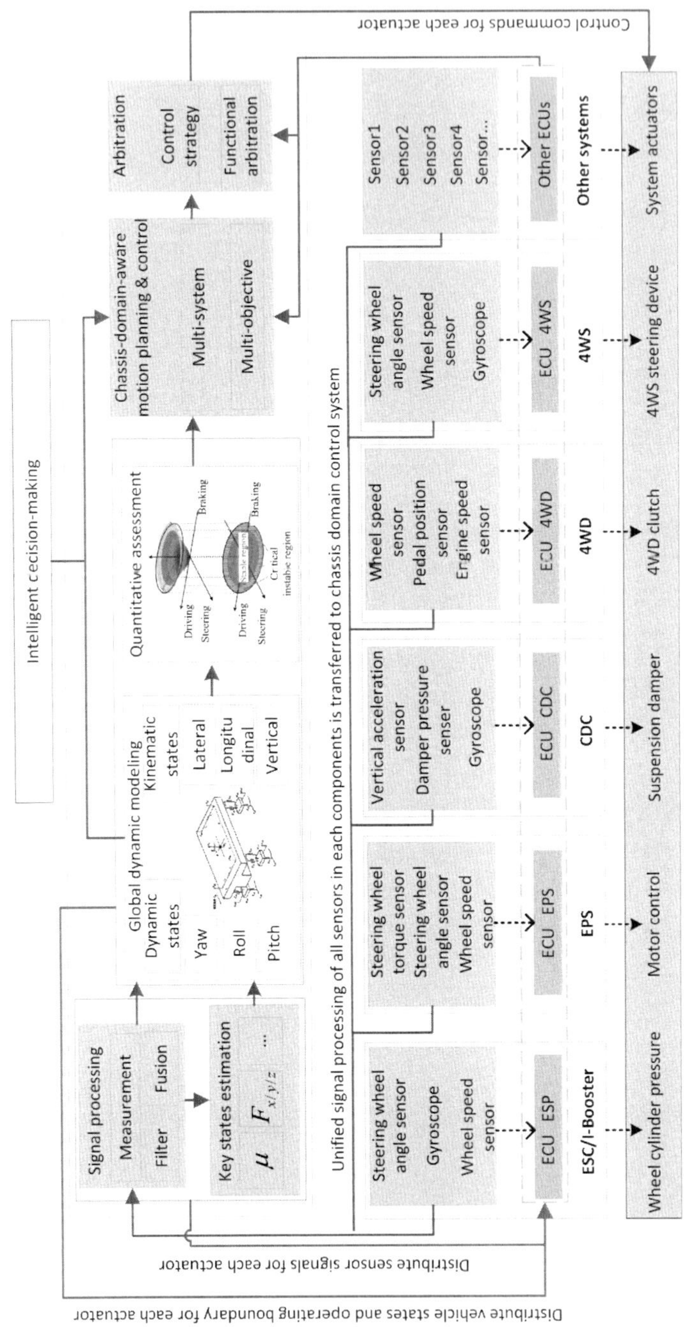

Figure 2.5 Concept of chassis domain dynamic control

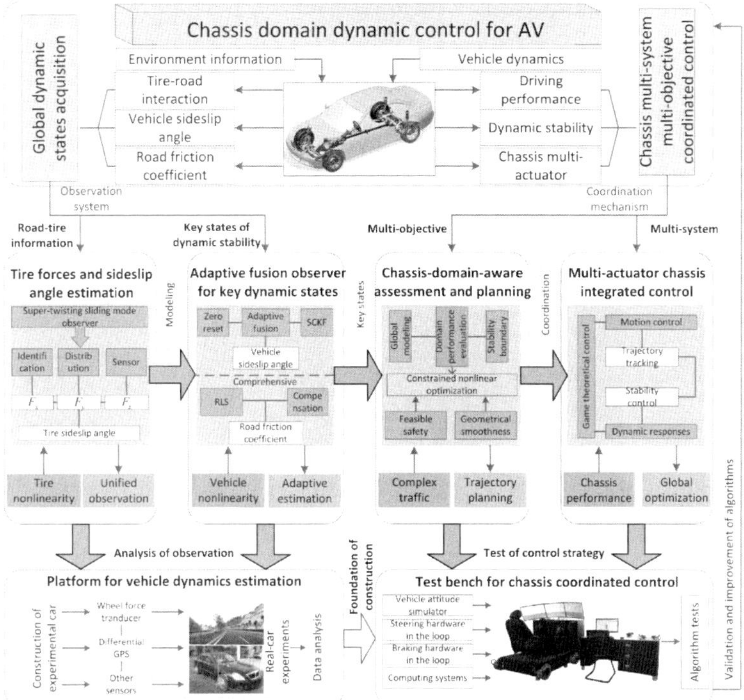

Figure 2.6 Overall framework of the chassis domain dynamic control.

The four submodules include a unified super-twisting sliding mode observer system for estimating tire forces and tire sideslip angles, an adaptive fusion observer for critical state estimation, a chassis-domain-aware assessment and motion planning submodule, and a multi-actuator integrated control submodule. The following will briefly describe the four submodules and the two verification platforms.

VIGNETTE 2.1 Observation System for Road-Tire Information

Accurate and unified observation of tire forces and tire sideslip angles is fundamental to global vehicle dynamic modeling. The critical challenge addressed in this book is the precise observation of individual tire forces and the accurate estimation of tire sideslip angles by utilizing onboard sensor signals. Additionally, the proposed observation system takes into account the impact of tire nonlinearity on observation accuracy. The unified super-twisting SMO (ST-SMO) system for estimating tire forces and sideslip angles employs the signals of onboard sensors. The first step is to design the approach for calculating vertical tire forces. Based on the

ST-SMO methodology, we develop a specific ST-SMO for longitudinal tire forces based on the identification of effective tire radius and vehicle longitudinal dynamics. Then, a specific ST-SMO for lateral tire forces and an ST-SMO dedicated to estimating tire sideslip angles are developed.

VIGNETTE 2.2 Adaptive Fusion Observer for Key Dynamic States

Critical dynamic stability states, especially vehicle sideslip angle, are pivotal for stability performance and control. Accurate observation of these key states, which are challenged by vehicle nonlinear dynamics, especially in conditions of extreme maneuvers, is a significant focus of this submodule. In this submodule, we first employ the CKF algorithm and develop an adaptive fusion method based on the CKF and zero-reset integration to achieve precise observation of the vehicle sideslip angle through the adaptive fusion of the two methods. Moreover, this submodule consists of a mode-switching observer which aims to estimate road friction coefficient. The mode-switching observer uses the comprehensive combination of an RLS method and the compensation algorithm to obtain the estimation value of the road friction coefficient.

VIGNETTE 2.3 Chassis-Domain-Aware Assessment and Motion Planning

In this submodule, we reveal mathematically explicit chassis steady boundaries of various vehicle maneuvers based on the modeling of chassis-domain dynamics and vehicle spatiotemporal signal analysis for safety-critical AVs. Then, a rigorous metric is developed to quantify vehicle chassis performance comprehensively. Based on the comprehensive performance metric, we develop the chassis-domain-aware motion planning method. Thereby, this submodule can perform the motion planning task and generate an optimal and dynamically feasible trajectory satisfying geometrical smoothness, chassis global dynamic requirements, and feasible safety.

VIGNETTE 2.4 Multi-Actuator Chassis Integrated Control

It is significant to operate accurately and stably the trajectory tracking task based on the desired trajectory generated by the motion planning submodule. The high nonlinearity of chassis dynamics and multi-actuator integrated control underlying this task is intractable, which involves the optimization of multiple performance targets, including trajectory tracking and handling stability. This book defines each wheel to be fully adjustable and controllable to overcome the individual actuation limitation of traditional chassis structures. A game-theoretical control scheme is proposed in this submodule to formulate chassis multi-actuator integrated control, and the chassis global dynamics can be optimized by guaranteeing a Nash equilibrium for this game.

VIGNETTE 2.5 Construction of Experimental Platforms

To validate and iteratively optimize the chassis domain system and implementations of each submodule, two platforms are designed and constructed: a dynamic observation platform adaptable to full conditions and a chassis-coordinated control test bench. The full-condition dynamic observation platform is built on an experimental Sedan car equipped with WFTs, double-antenna differential GPS, and some other sensors. It is employed to conduct experiments on the observations and dynamic modeling. The chassis-coordinated control test bench is constructed based on actual vehicle configuration parameters. The chassis test bench includes a vehicle attitude simulator, hardware-in-the-loop setups for steering and braking systems, and upper and lower computing systems. We carry out algorithm tests and validations of the chassis integrated control methods on this chassis test bench.

2.4 ANALYSIS OF DYNAMIC MODELING

A vehicle is a complex system comprising multiple dynamic components such as the engine system, braking system, steering system, wheels, suspension, and vehicle body. Constructing an accurate and complete model for such a system poses significant challenges. However, a vehicle dynamic model is fundamental for conducting vehicle dynamic analysis and developing control strategies. To facilitate theoretical dynamic analysis and control strategy research, this section will provide some explanation about the tire dynamics model and vehicle dynamic model.

2.4.1 ANALYSIS OF TIRE DYNAMICS

To describe the mechanical characteristics of a tire, the tire coordinate system is defined as shown in Fig. 2.7, with the origin at the tire print center. The x-axis is aligned with the line of the intersection between the wheel plane and the road surface. The z-axis is perpendicular to the road plane. The y-axis lies in the road plane. A tire with a rolling radius R has dynamic parameters: tire

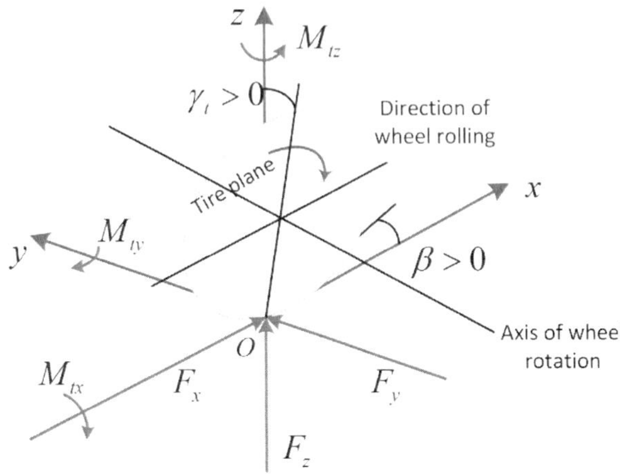

Figure 2.7 Demonstration of tire coordinate system and tire forces.

forces F_x, F_y, and F_z along the x, y, and z axes, respectively, and moments of force M_{tx}, M_{ty}, and M_{tz} around these axes. The primary input states of a tire include the tire angular speed ω, longitudinal velocity v_{tx}, lateral velocity v_{ty}, and tire camber angle γ_t. Additionally, parameters related to vehicle dynamic stability are the tire sideslip angle α and tire slip ratio λ. The tire sideslip angle is the angle between the velocity vector at the tire rotational center and the tire's geometrical central plane, which is calculated by Eq. (2.1). The tire slip ratio is calculated by Eq. (2.2).

$$\alpha = \arctan\left(\frac{v_{ty}}{v_{tx}}\right) \tag{2.1}$$

$$\lambda = \frac{\omega R - v_{tx}}{\max(v_{tx}, \omega R)} \tag{2.2}$$

Common tire models are the linear tire model, Dugoff tire model, and magic formula tire model, which are described in the following.

VIGNETTE 2.6 Linear Tire Model

The linear tire model is typically used for analyzing the steady-state steering characteristics of a vehicle [51]. In the situations of small sideslip angles, the relationship between the lateral force and tire sideslip angle can be depicted as a linear relationship, as shown in Eq. (2.3), where F_y represents tire lateral force, α is the sideslip angle, and k_α is the cornering stiffness, defined as the slope of the curve that depicting tire lateral force versus sideslip angle when the sideslip angle is near zero.

$$F_y = k_\alpha \cdot \alpha \tag{2.3}$$

VIGNETTE 2.7 Dugoff Tire Model

The Dugoff tire model is expressed by Eqs. (2.4), (2.5), and (2.6) [40]. In these equations, μ and μ_p represent the road adhesion coefficient and the peak adhesion coefficient, respectively. A_s denotes the shape factor. C_α and C_λ are the tire cornering stiffness and slip stiffness, and F_N represents the tire vertical load.

$$F_x = \begin{cases} \dfrac{\lambda}{1-\lambda} \cdot C_\lambda, H < \frac{1}{2}, \\ \dfrac{\lambda}{1-\lambda} \cdot C_\lambda \left(\dfrac{1}{H} - \dfrac{1}{4H^2} \right), H \geq \frac{1}{2}, \end{cases} \tag{2.4}$$

$$F_y = \begin{cases} \dfrac{\lambda}{1-\lambda} \cdot C_\alpha \cdot \alpha, H < \frac{1}{2}, \\ \dfrac{\lambda}{1-\lambda} \cdot C_\alpha \left(\dfrac{1}{H} - \dfrac{1}{4H^2} \right) \cdot \tan\alpha, H \geq \frac{1}{2}, \end{cases} \tag{2.5}$$

$$\begin{cases} H = \sqrt{\left(\dfrac{1}{1-\lambda} \cdot \dfrac{C_\lambda}{\mu F_N} \right)^2 + \left(\dfrac{1}{1-\lambda} \cdot \dfrac{C_\alpha}{\mu F_N} \cdot \tan\alpha \right)^2} \\ \mu = \mu_p (1 - A_s \cdot R \cdot \omega \sqrt{\lambda^2 + \tan^2\alpha}) \\ \tan\alpha \approx \alpha = \delta_f - \beta \end{cases} \tag{2.6}$$

VIGNETTE 2.8 Magic Formula Tire Model

Pacejka's magic formula model [106] can be expressed as Eq. (2.7), where $Y(x)$ represents the tire longitudinal force or lateral force, and x denotes the longitudinal slip ratio or tire sideslip angle.

$$Y(x) = D \sin\{C \arctan[B(x + s_h) - E(B(x + s_h) - \arctan(B(x + s_h)))]\} + s_v \tag{2.7}$$

When we use Eq. (2.7) to calculate the longitudinal tire forces, these parameters can be calculated by

$$\begin{cases} C = b_0, \\ D = b_1 \cdot F_z^2 + b_2 \cdot F_z, \\ B = \dfrac{b_3 \cdot F_z^2 + b_4 \cdot F_z}{C \cdot D \cdot e^{b_5 F_z}}, \\ E = b_6 \cdot F_z^2 + b_7 \cdot F_z + b_8 \end{cases} \tag{2.8}$$

If the magic formula model is utilized to obtain lateral tire forces, we should use the following equation to calculate those parameters.

$$\begin{cases} C = a_0, \\ D = \mu\left(a_1 \cdot F_z^2 + a_2 \cdot F_z\right), \\ B = a_3 \sin[2\arctan(F_z/a_4)]/(C \cdot D), \\ E = a_6 \cdot F_z + a_7 \end{cases} \tag{2.9}$$

where all the hyperparameterss$_h$, s_v, $b_0 - b_8$, and $a_0 - a_7$ can be calibrated through tire tests.

Considering the constraints of the tire friction ellipse, we have to modify the magic formula model by

$$\begin{cases} F_x = \dfrac{|\sigma_x|}{\sigma} Y(\lambda), \\ F_y = \dfrac{|\sigma_y|}{\sigma} Y(\alpha), \\ \sigma = \sqrt{\sigma_x^2 + \sigma_y^2}, \\ \sigma_x = \dfrac{\lambda}{1+\lambda}, \\ \sigma_y = \dfrac{\tan\alpha}{1+\lambda} \end{cases} \tag{2.10}$$

2.4.2 ANALYSIS OF VEHICLE DYNAMICS

To investigate related issues on chassis dynamic control, it is essential to simplify the vehicle's complex physical system to a certain extent. In this chapter, we briefly explain two universal models.

VIGNETTE 2.9 Four-Wheel 7DOF Dynamic Modeling

As shown in Fig. 2.8, the four-wheel 7DOF model ignores the effects of vehicle rolling, pitching, and vertical motion. In addition, the influences of aerodynamic drag, rolling resistance, steering system, and suspension system are not considered as well. It describes the following vehicle motions: longitudinal motion, yaw, lateral motion, and the rotational motion of each wheel. This model comprehensively reflects the dynamic characteristics related to vehicle dynamic stability, thus it has been used extensively to conduct vehicle dynamic stability analysis and control. The mathematical description of this model is as follows. Vehicle longitudinal motion:

$$m(\dot{v}_x - v_y \varphi) = F_{xrl} + F_{xrr} - (F_{yfl} + F_{yfr})\sin\delta + (F_{xfl} + F_{xfr})\cos\delta \tag{2.11}$$

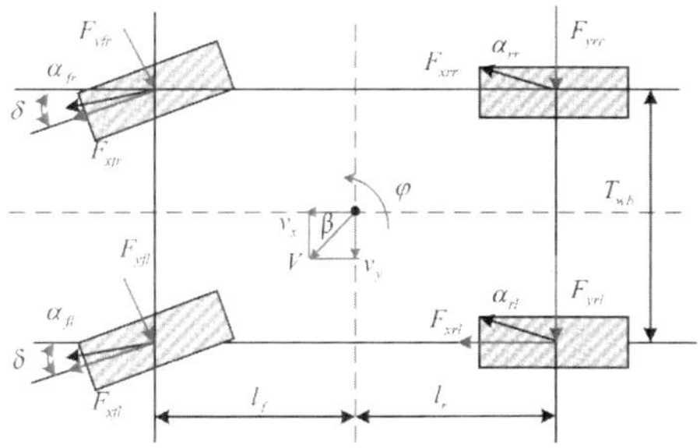

Figure 2.8 Demonstration of four-wheel 7DOF dynamic model.

Vehicle lateral motion:

$$mv_x(\dot{\beta}+\varphi) = F_{yrl} + F_{yrr} + (F_{yfl} + F_{yfr})\cos\delta + (F_{xfl} + F_{xfr})\sin\delta \quad (2.12)$$

Vehicle yaw motion:

$$I_z\dot{\varphi} = (F_{yfl} + F_{yfr})l_f \cdot \cos\delta + (F_{yfl} - F_{yfr})\frac{T_{wb}}{2}\sin\delta - (F_{yrl} + F_{yrr})l_r + (F_{xfl} + F_{xfr})l_f \cdot \sin\delta - (F_{xfl} - F_{xfr})\frac{T_{wb}}{2}\cos\delta - (F_{xrl} - F_{xrr})\frac{T_{wb}}{2} \quad (2.13)$$

Rotational dynamic equation for each wheel:

$$J_w\dot{\omega}_{ij} = T_{wtij} - T_{wpij} - F_{xij}R \quad (2.14)$$

VIGNETTE 2.10 Single-Track Dynamic Modeling

Based on the four-wheel 7DOF model, we can further simplify and derive the single-track 2DOF model. As illustrated in Fig. 2.9, we first neglect four-wheel rotational motions and then simplify vehicle longitudinal motion by considering vehicle longitudinal speed as an external input. Additionally, we assume complete symmetry between the left and right

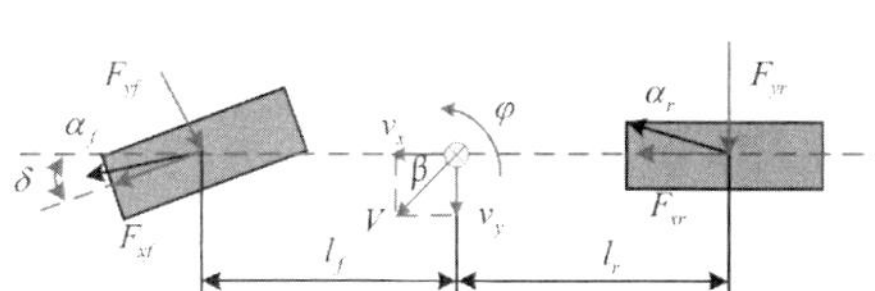

Figure 2.9 Demonstration of the single-track dynamic model.

wheels. Thus, we can get the single-track dynamic model, which is the most fundamental vehicle dynamic model. It only contains the vehicle lateral sideslip angle and vehicle yaw rate, reflecting vehicle lateral and yaw motion characteristics during steering maneuvers. This model can reveal the relationship between vehicle handling stability and key vehicle dynamic states, i.e., vehicle yaw rate and vehicle sideslip angle. The mathematical description of this model is given by

$$m(\dot{v}_y + v_x\varphi) = F_{yf} + F_{yr} \tag{2.15}$$

$$I_z\dot{\varphi} = l_f F_{yf} - l_r F_{yr} \tag{2.16}$$

The sideslip angle of the front and rear axles can be calculated by

$$\begin{cases} \alpha_f = \delta - (v_y + l_f\varphi)/v_x \\ \alpha_r = -(v_y - l_r\varphi)/v_x \end{cases} \tag{2.17}$$

If we combine Eq. (2.17) and the linear tire model in Eq. (2.3) with the above equations of the single-track dynamic model, the following equation can be yielded:

$$\begin{cases} (k_f + k_r)\beta + \dfrac{1}{v_x}(l_f k_f - l_r k_r)\varphi - k_f\delta = m(\dot{v}_y + v_x\varphi) \\ (l_f k_f - l_r k_r)\beta + \dfrac{1}{v_x}(l_f^2 k_f + l_r^2 k_r)\varphi - l_f k_f\delta = I_z\dot{\varphi} \end{cases} \tag{2.18}$$

2.5 CONCLUSION

This chapter first introduces the research background of chassis dynamic integrated control and summarizes three main architectures. Then, we present the analysis and design of the overall architecture of the chassis dynamic domain control for AVs. It includes four submodules within the overall architecture, and this chapter provides a brief explanation and overview implementation of each submodule involved in the architecture. Tire dynamics and vehicle dynamics are analyzed, and common models are presented as well.

3 Identification of Tire–Road Interactions

This chapter focuses on estimating tire dynamic parameters, including the vertical, longitudinal, and lateral tire forces for all wheels and the tire sideslip angle. A literature review summarizing recent investigations on tire dynamics observation is provided. This chapter then proposes a unified super-twisting SMO method and derives specific observers for each component of tire dynamic states. A comprehensive dynamic observation platform is established to verify the observation algorithms, equipped with WFTs for tires, dual-axis optical sensors for measuring tire sideslip angle, and onboard sensors.

3.1 BACKGROUND

Knowledge of tire–road information is crucial not only in many active safety applications but also for self-driving cars. Interesting investigations on vehicle tire dynamics estimation have been conducted. For example, L. R. Ray proposed an EKF-based observer to estimate the state and longitudinal and lateral tire force histories of a nine-degree-of-freedom vehicle [114]. Then, the EKF was utilized as a robust tire force estimator, which was validated against results from strain-gauged wheel rims [140]. Afterward, many EKF-based tire force estimators were designed [9, 34, 116]. Two observers based on the EKF and UKF were proposed by [37], and experimental results demonstrated that UKF is a superior alternative, especially when the system presents strong nonlinearities. Some tire-force estimation algorithms have been applied to vehicle dynamics control. For example, T. Hsiao proposed an observer-based traction force control scheme and it could be robust with respect to variations in road conditions and tire model uncertainties [59]. W. Cho et al. designed a scheme for longitudinal/lateral tire-force estimation and it was integrated into a unified chassis control system, i.e. vehicle stability control [31].

Though many tire-force estimators so far have been designed, their performance still needs to be improved. When the vehicle is in a non-linear state, EKF-based estimators' accuracy decreases because of the linearization method utilized in EKF which only ensures first-order accuracy [18]. UKF-based estimators' stability requires the prediction matrix to be positive definite [13]. Various road conditions cause uncertainty fluctuation of tire–road friction coefficient, so its observation still needs to be further researched. Research about the observation of tire–road information is still insufficient.

DOI: 10.1201/9781003583929-3

3.2 DESIGN OF TIRE DYNAMIC PARAMETER ESTIMATION ALGORITHMS

This chapter uses the four-wheel 7DOF dynamic model, which is introduced in Section 2.4, to develop the estimation method for tire dynamic parameters. The conventional onboard sensors used in the estimation of tire dynamic parameters include the steering angle sensor, wheel speed sensor, and wheel cylinder pressure sensor. This chapter introduces the framework of the proposed ST-SMO, as shown in Fig. 3.1. It includes the vertical tire force calculation module, the longitudinal force observation module based on the identification of effective tire radius, the lateral tire force observation module, the tire sideslip angle observation module, and the tire sideslip angle correction component.

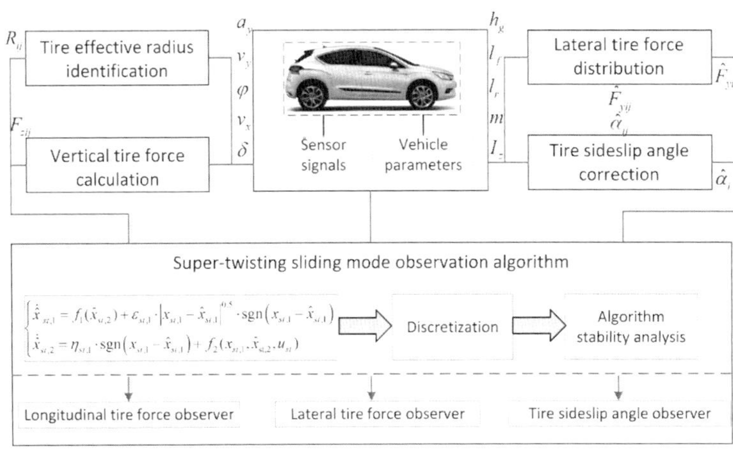

Figure 3.1 Adaptive unified observation of four-wheel tire force and tire sideslip angle.

3.2.1 VERTICAL TIRE FORCE CALCULATION

The distribution of vertical tire forces during vehicle maneuvers is not only influenced by static factors such as the vehicle mass and structural layout of the vehicle body but also closely related to the dynamic responses. The longitudinal acceleration or deceleration can cause changes in vehicle pitching, and steering can induce changes in lateral acceleration and vehicle rolling. These factors affect the vertical forces on each tire, as illustrated in Fig. 3.2.

When the vehicle is stationary, the static vertical tire forces can be calculated by leveraging the equilibrium of forces and moments acting on the vehicle, as depicted in Eq. (3.1). In this situation, the vertical tire forces are

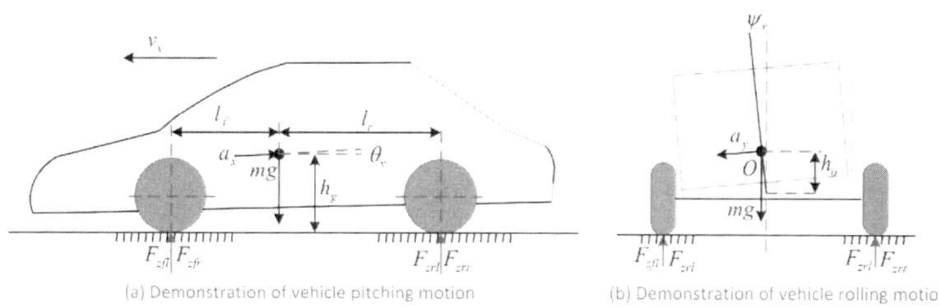

Figure 3.2 Demonstrations of vehicle pitching and rolling motions.

equal on the left and right wheels, i.e., $F^s_{zfl} = F^s_{zfr}$ and $F^s_{zrl} = F^s_{zrr}$.

$$\begin{cases} (F^s_{zfl} + F^s_{zfr})(l_f + l_r) = m \cdot g \cdot l_r \\ (F^s_{zrl} + F^s_{zrr})(l_f + l_r) = m \cdot g \cdot l_f \end{cases} \tag{3.1}$$

As depicted in Fig. 3.2 (a), an inertial force caused by the longitudinal acceleration a_x will be applied to the vehicle sprung mass, resulting in pitching motion with a pitching angle θ_v, when the vehicle operates acceleration or deceleration maneuvers. Eq. (3.1) represents the force equilibrium equation for a stationary vehicle. The vertical tire forces at each tire print and the gravitational force acting on the vehicle mass form a balanced force system. By isolating this system, we can independently analyze the vertical tire forces for the front and rear wheels in the cases of longitudinal acceleration. Moreover, the vertical tire forces on the left and right wheels are equal. Applying torque balance equations separately for the front and rear wheels, we can yield:

$$\begin{cases} (F^p_{zfl} + F^p_{zfr})(l_f + l_r) = -h_g \cdot m \cdot a_x \cdot \cos\theta_v + l_r \cdot m \cdot a_x \cdot \sin\theta_v \\ (F^p_{zrl} + F^p_{zrr})(l_f + l_r) = h_g \cdot m \cdot a_x \cdot \cos\theta_v + l_f \cdot m \cdot a_x \cdot \sin\theta_v \end{cases} \tag{3.2}$$

In addition to the changes in vertical tire forces caused by the vehicle mass and longitudinal acceleration, lateral acceleration and the rolling motion of the sprung mass during vehicle steering maneuvers also cause axle load transfer, as illustrated in Fig. 3.2 (b). Considering the effects of lateral acceleration and vehicle rolling motion, we can similarly derive the following:

$$\begin{cases} F^r_{zfl} = -\dfrac{l_r m h_g}{T_{wb}(l_f + l_r)} a_y \cos\psi_r - \dfrac{l_r(0.5T_{wb} + h_o \sin\psi_r)}{T_{wb}(l_f + l_r)} m a_y \sin\psi_r \\ F^r_{zfr} = \dfrac{l_r m h_g}{T_{wb}(l_f + l_r)} a_y \cos\psi_r - \dfrac{l_r(0.5T_{wb} - h_o \sin\psi_r)}{T_{wb}(l_f + l_r)} m a_y \sin\psi_r \\ F^r_{zrl} = -\dfrac{l_f m h_g}{T_{wb}(l_f + l_r)} a_y \cos\psi_r - \dfrac{l_f(0.5T_{wb} + h_o \sin\psi_r)}{T_{wb}(l_f + l_r)} m a_y \sin\psi_r \\ F^r_{zrr} = \dfrac{l_f m h_g}{T_{wb}(l_f + l_r)} a_y \cos\psi_r - \dfrac{l_f(0.5T_{wb} - h_o \sin\psi_r)}{T_{wb}(l_f + l_r)} m a_y \sin\psi_r \end{cases} \tag{3.3}$$

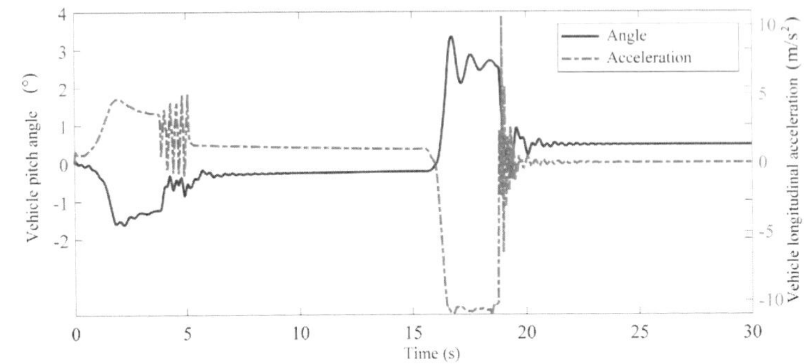

Figure 3.3 Vehicle pitching angle and longitudinal acceleration curves.

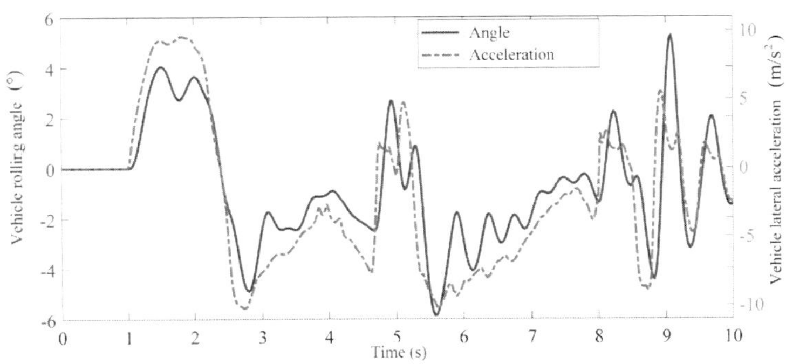

Figure 3.4 Vehicle rolling angle and lateral acceleration curves.

where h_o denotes the distance between the center of the vehicle sprung mass and the vehicle rolling center.

Typically, the pitching stiffness of passenger cars ranges from 80000 to 140000 $N \cdot m/rad$, the front-axle rolling stiffness ranges from 17000 to 69000 $N \cdot m/rad$, the rear-axle rolling stiffness ranges from 10000 to 40000 $N \cdot m/rad$, the front-rolling center height of the vehicle ranges from 0 to 0.14 m, and the rear rolling center height ranges from 0 to 0.40 m. These specific parameters are tested, designed, and calibrated via vehicle K&C tests to tune vehicle chassis performance. The curves of vehicle pitching and rolling response in extreme maneuvers, such as high acceleration, emergency braking, and large-angle steering, are depicted in Figs. 3.3 and 3.4.

We will analyze vehicle dynamic responses as demonstrated in Figs. 3.3 and 3.4. The car performs a high acceleration maneuver with a longitudinal acceleration reaching 6 m/s^2 in the beginning 5 s. During this acceleration maneuver, the vehicle's peak pitching angle is approximately $-1.8°$. Subsequently, the vehicle accelerates smoothly during the next 10 s with a longitudinal acceleration of about 0.5 m/s^2, and the vehicle pitching angle remains stable at about $-0.2°$. From about 16 s to 19 s, the vehicle performs an emergency braking with a deceleration of -10 m/s^2. During this procedure, the peak value of the vehicle pitching angle is $3°$, varying within the range from $2°$ to $3°$. Fig. 3.4 depicts vehicle dynamic responses during the fishhook maneuver, including lateral acceleration and rolling angle. The variation trend of the lateral acceleration curve is similar to that of the rolling angle curve. The peak value of lateral acceleration approaches 10 m/s^2, and the corresponding peak rolling angle reaches $5°$. This analysis shows strong evidence that in normal driving cases and even in extreme maneuvering conditions, the pitching motion of vehicle sprung mass remains within $5°$, and the rolling angle stays within $6°$. Hence, we can yield $\cos\theta_v \approx 1$, $\cos\psi_r \approx 1$, $\sin\theta_v \approx 0$, and $\sin\psi_r \approx 0$. Additionally, mass-produced vehicles usually do not have sensors, especially IMU for the real-time measurement of vehicle sprung mass rolling and pitching motions. Fortunately, common cars can measure vehicle longitudinal acceleration, lateral acceleration, and yaw rate Thus, this chapter ignores minor factors and considers only the effects of longitudinal and lateral accelerations on the vertical load of each tire. The calculation method for tire vertical forces applicable to mass-produced vehicles can be described as shown in Eq. (3.4).

$$\begin{cases} F_{zfl} = mgl_r/2(l_f+l_r) - ma_xh_g/2(l_f+l_r) - ml_rh_gv_x\varphi/T_{wb}(l_f+l_r) \\ F_{zfr} = mgl_r/2(l_f+l_r) - ma_xh_g/2(l_f+l_r) + ml_rh_gv_x\varphi/T_{wb}(l_f+l_r) \\ F_{zrl} = mgl_f/2(l_f+l_r) + ma_xh_g/2(l_f+l_r) - ml_fh_gv_x\varphi/T_{wb}(l_f+l_r) \\ F_{zrr} = mgl_f/2(l_f+l_r) + ma_xh_g/2(l_f+l_r) + ml_fh_gv_x\varphi/T_{wb}(l_f+l_r) \end{cases} \tag{3.4}$$

3.2.2 SLIDING MODE OBSERVERS

3.2.2.1 Basic Issues of Sliding Mode Variable Structure Theory

Sliding Mode Control (SMC), as the name suggests, involves a control system with a "structure" that is not fixed. In an SMC system, we need to first design a so-called "sliding mode," which allows the controlled system states to move along this sliding mode to achieve the desired target. This process requires continuous adjustments based on the changing state of the controlled system, for instance, the state deviation or its derivatives [109]. This method was first proposed and designed by Soviet experts Utkin and Emelyanov. Over the past 70 years, it has been rapidly developed and widely applied in various aspects of control engineering: from various types of linear systems to nonlinear systems, from deterministic systems to uncertain systems, and it can be applied not only to discrete systems but also to continuous systems, forming numerous research branches [123]. It has gradually been applied in various engineering practices, such as motor state observation and control [153], vehicle chassis component parameter observation and electronic control [65], and satellite attitude control [42]. We do not have to know exactly the parameters of the controlled object or disturbances when designing the sliding mode. In other words, SMC has strong robustness, which is one of its significant advantages. Additionally, the implementation of SMC is straightforward, and the algorithm's response speed is extremely fast. However, traditional SMC methods also face a tricky problem known as "chattering." Concretely, discontinuous functions make it difficult for the system trajectory to slide along the sliding mode to the equilibrium point without error, which causes the system to oscillate back and forth after reaching the sliding surface.

For a general n-dimensional system of the form $\dot{\boldsymbol{x}}_{smc} = f(\boldsymbol{x}_{smc})$, where $\boldsymbol{x}_{smc} \in \mathbb{R}^n$, there exists a switching surface $S(x_{smc,1}, x_{smc,2}, \ldots, x_{smc,n}) = 0$ in the state space that divides the state space into two parts: $S(\boldsymbol{x}_{smc}) > 0$ and $S(\boldsymbol{x}_{smc}) < 0$.

As shown in Fig. 3.5, based on the trajectory variation around the switching surface, the system motion can be classified into the following three situations: regular point, starting point, and ending point. When the controlled system trajectory approaches the switching surface and then directly crosses it, we call this point a regular point, which is denoted by point A as shown in Fig. 3.5. When the state trajectory near the switching surface leaves the switching surface from both sides, we call this point as a starting point, which is denoted by the point B in Fig. 3.5. When the trajectory of the controlled object reaches the vicinity of the switching surface and gradually approaches it from both sides, it is called an ending point, as shown by point C in Fig. 3.5. SMC requires that the initial state of the controlled object should approach the switching surface when it is not around the switching surface, i.e., the system's sliding mode should be reachable. When the system's sliding surface satisfies Eq. (3.5), its trajectory can move to the vicinity of the switching surface within a certain period, and after entering the sliding mode, the trajectory of the controlled object can be maintained around the switching surface. In practical applications, chattering phenomena occur due to the inherent characteristics of discrete systems, the time and spatial lag of switches, and the system inertia.

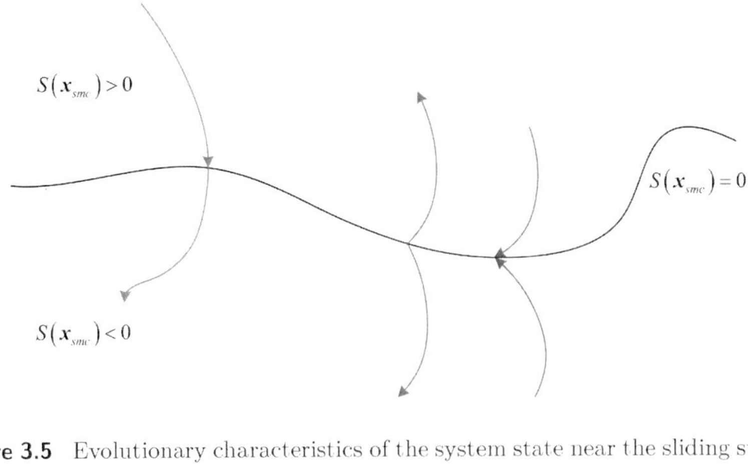

Figure 3.5 Evolutionary characteristics of the system state near the sliding surface.

$$S(\boldsymbol{x}_{smc})\dot{S}(\boldsymbol{x}_{smc}) \leq 0 \Leftrightarrow \begin{cases} \lim\limits_{S\to+0} \dot{S}(\boldsymbol{x}_{smc}) \leq 0 \\ \lim\limits_{S\to-0} \dot{S}(\boldsymbol{x}_{smc}) \geq 0 \end{cases} \tag{3.5}$$

3.2.2.2 Design of Super-Twisting Sliding Mode Observers

On the one hand, observers based on SMC theory can adjust the state trajectory according to the desired sliding-mode path. Moreover, system disturbances do not affect the observation of the system states which have reached the sliding mode. Therefore, the advantage of the SMO algorithms lies in their fast dynamic response and strong robustness against system disturbances. On the other hand, considering the expression forms of the vehicle dynamic model, we can derive unified mathematical expressions for longitudinal tire force, lateral tire force, and tire sideslip angle (the detailed derivation will be given later). This chapter proposes the observers for tire forces and tire sideslip angles of each wheel based on the SMO algorithm. To mitigate the inherent chattering phenomenon that limits the application of SMOs, a new unified SMO system, named ST-SMO, is proposed for the unified, precise, and robust observation of tire dynamic parameters.

The ST-SMO algorithm is a second-order sliding-mode algorithm. Let us consider a general n-dimensional system as shown in Eq. (3.6), and its state space is given by $\boldsymbol{x}_{st,n} = [x_{st,1}, x_{st,2}, \ldots, x_{st,n}]$.

$$\begin{cases} \dot{\boldsymbol{x}}_{st,n-1} = \begin{bmatrix} 0 & 1 & 0 & 0 & 0 \\ 0 & 0 & 1 & 0 & 0 \\ & & & \ddots & \\ 0 & 0 & 0 & \ddots & 0 \\ 0 & 0 & 0 & 0 & 1 \end{bmatrix} \boldsymbol{x}_{st,n} \\ \dot{x}_{st,n} = f(x_{st,1}, x_{st,2}, \ldots, x_{st,n-1}) \\ y_{st} = x_{st,1} \end{cases} \tag{3.6}$$

Based on the super-twisting theory [124], the proposed ST-SMO for the system described by Eq. (3.6) can be designed as follows:

$$\begin{cases} \dot{\hat{x}}_{st,1} = \tilde{x}_{st,2} + \varepsilon_{st,1} \cdot |x_{st,1} - \hat{x}_{st,1}|^{0.5} \cdot \mathrm{sgn}\,(x_{st,1} - \hat{x}_{st,1}) \\ \dot{\tilde{x}}_{st,2} = \eta_{st,1} \cdot \mathrm{sgn}\,(x_{st,1} - \hat{x}_{st,1}) \\ \dot{\hat{x}}_{st,2} = E_{st,1} \cdot \left[\tilde{x}_{st,3} + \varepsilon_{st,2} \cdot |\tilde{x}_{st,2} - \hat{x}_{st,2}|^{0.5} \cdot \mathrm{sgn}\,(\tilde{x}_{st,2} - \hat{x}_{st,2})\right] \\ \quad \vdots \\ \dot{\tilde{x}}_{st,n} = E_{st,n-2} \cdot \eta_{st,n-1} \cdot \mathrm{sgn}\,(\tilde{x}_{st,n-1} - \hat{x}_{st,n-1}) \\ \dot{\hat{x}}_{st,n} = E_{st,n-1} \cdot \left[\tilde{\theta}_{st} + \varepsilon_{st,n} \cdot |\tilde{x}_{st,n} - \hat{x}_{st,n}|^{0.5} \cdot \mathrm{sgn}\,(\tilde{x}_{st,n} - \hat{x}_{st,n})\right] \\ \tilde{\theta}_{st} = E_{st,n-1} \cdot \eta_{st,n} \cdot \mathrm{sgn}\,(\tilde{x}_{st,n} - \hat{x}_{st,n}) \end{cases}$$
$$E_{st,i} = \begin{cases} 1 & \tilde{x}_{st,i} - \hat{x}_{st,i} < \sigma_{st} \\ 0 & else \end{cases} \tag{3.7}$$

where $\varepsilon_{st,1} \sim \varepsilon_{st,n}$ and $\eta_{st,1} \sim \eta_{st,n}$ are the gain parameters of the ST-SMO algorithm. σ_{st} is a small positive real number.

In this book, we focus on the observation of vehicle tire longitudinal force, lateral force, and tire sideslip angle. Therefore, based on the mathematical expressions derived from the vehicle dynamic model (which will be elaborated on later), an ST-SMO method for the system with two state variables is designed as follows:

$$\begin{cases} \dot{\hat{x}}_{st,1} = f_1(\hat{x}_{st,2}) + \varepsilon_{st,1} \cdot |x_{st,1} - \hat{x}_{st,1}|^{0.5} \cdot \operatorname{sgn}(x_{st,1} - \hat{x}_{st,1}) \\ \dot{\hat{x}}_{st,2} = \eta_{st,1} \cdot \operatorname{sgn}(x_{st,1} - \hat{x}_{st,1}) + f_2(x_{st,1}, \hat{x}_{st,2}, u_{st}) \end{cases} \quad (3.8)$$

where u_{st} is the system control input.

Introducing the ST-SMO algorithm as shown in Eq. (3.8) into the specific design of the observer for longitudinal tire force, lateral tire force, and tire sideslip angle, we can establish a unified framework based on the ST-SMO algorithm for the estimation of tire dynamic parameters.

3.2.3 OBSERVATION OF LONGITUDINAL TIRE FORCE

Because the effective tire radius is an important parameter of tire characteristics and significantly affects tire dynamic responses, especially the longitudinal tire force, this book designs a longitudinal tire force observer based on the identification of the effective tire radius. Generally, the effective tire radius depends on key factors such as the vehicle tire type, the tire stiffness, the vehicle load, and the wheel speed. Fig. 3.6 illustrates the tire effective radius R_{ij}, the tire radius in an unloaded situation $R_{u,ij}$, and the static load radius $R_{s,ij}$. The effective tire radius is defined as the ratio of the tire's linear velocity to the wheel speed of that tire. The radius of the tire in a static, unloaded state is the unloaded tire radius. When the vehicle is stationary, the distance from the tire center to the tire print center on the ground is the tire's static

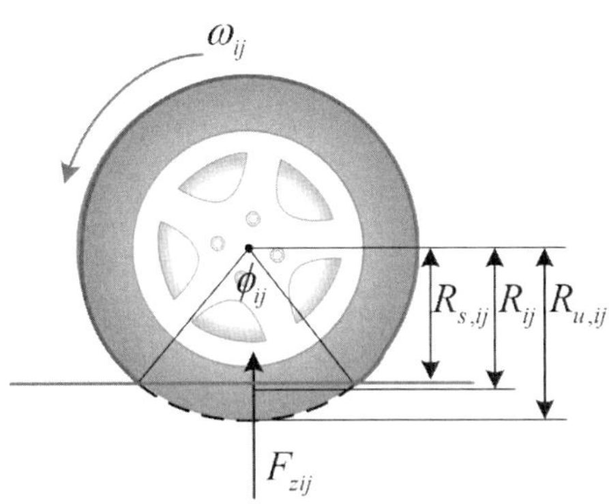

Figure 3.6 Illustration of the tire effective radius.

load radius. However, the effective tire radius is not equal to the static load radius, rather, it is related to both the static load radius and the unloaded tire radius, satisfying $R_{s,ij} < R_{ij} < R_{u,ij}$ [62].

When a wheel rolls forward on the road, the tire circumferential part passing through the tire print flattens out. Therefore, the effective tire radius can be estimated by replacing the tire circumferential part with a straight line corresponding to the tire print. As shown in Fig. 3.6, the tire vertical deflection can be calculated by the following equation:

$$R_{u,ij} - R_{s,ij} = R_{u,ij}(1 - \cos\frac{\phi_{t,ij}}{2}) \tag{3.9}$$

At the same time, by analyzing the geometric model at the tire print, we can yield:

$$R_{s,ij} = R_{u,ij}\cos\frac{\phi_{t,ij}}{2} \tag{3.10}$$

$$\frac{p_{len}}{2} = R_{u,ij}\sin\frac{\phi_{t,ij}}{2} \tag{3.11}$$

where p_{len} denotes the length of the tire print.

Let us Assume that the tire motion is equivalent to the rolling of a rigid body with a radius of R_{ij}, then when the tire rotational angle is $2\phi_{t,ij}$, its displacement is $p_{len} = R_{ij}\phi_{t,ij}$, that is:

$$p_{len} = 2R_{u,ij}\sin\frac{\phi_{t,ij}}{2} = R_{ij}\phi_{t,ij} \tag{3.12}$$

Thus, we can yield:

$$R_{ij} = \frac{R_{u,ij}\sin\frac{\phi_{t,ij}}{2}}{\frac{\phi_{t,ij}}{2}} \tag{3.13}$$

By employing the Taylor series expansion of the trigonometric functions and neglecting higher-order terms, we can obtain the following:

$$\frac{\sin\frac{\phi_{t,ij}}{2}}{\frac{\phi_{t,ij}}{2}} \approx 1 - \frac{1}{6}\left(\frac{\phi_{t,ij}}{2}\right)^2 \tag{3.14}$$

$$\cos\frac{\phi_{t,ij}}{2} \approx 1 - \frac{1}{2}\left(\frac{\phi_{t,ij}}{2}\right)^2 \Rightarrow \left(\frac{\phi_{t,ij}}{2}\right)^2 \approx 2\left(1 - \cos\frac{\phi_{t,ij}}{2}\right) \tag{3.15}$$

Combining Eq. (3.10), Eq. (3.13), and the Taylor series expansion of the trigonometric functions mentioned above, we can derive the following:

$$R_{ij} = \frac{2}{3}R_{s,ij} + \frac{1}{3}R_{u,ij} \tag{3.16}$$

The static load radius is a function of the tire vertical load, which is given by Eq. (3.17), where k_{zij} represents the vertical stiffness of each tire. Extensive experimental results show that the tire vertical stiffness is mainly influenced by tire pressure and the tire load. However, its value can be approximated as a constant at a given tire pressure [62]. For the tires of traditional passenger cars, this value can be approximated as 200 kN/m.

$$R_{s,ij} = R_{u,ij} - \frac{F_{zij}}{k_{zij}} \tag{3.17}$$

If we combine Eqs. (3.16) and (3.17), the formula for identifying the tire effective radius can be derived as follows:

$$R_{ij} = R_{s,ij} - \frac{1}{3}\frac{F_{zij}}{k_{zij}} \tag{3.18}$$

The rotational equation for each wheel in the four-wheel 7DOF dynamic model has been derived, as shown in Section 2.4, and it involves wheel acceleration $\dot{\omega}$, longitudinal tire force F_x, driving torque T_{wt}, and braking torque T_{wp}, in conjunction with the previously discussed tire effective radius identification method. Thus, we can conduct the design of the tire longitudinal force observation algorithm by further deriving the previously proposed ST-SMO method. Given that the wheel rotational speed can be measured, and the variable to be observed is the longitudinal tire force, this chapter defines that the state variables for the longitudinal tire force observer are ω and F_x. The construction of the ST-SMO algorithm for longitudinal tire force based on the identification of tire effective radius can be expressed by

$$\begin{cases} \dot{\hat{\omega}} = -\dfrac{R}{J_w} \cdot \hat{F}_x + \dfrac{T_{wt} - T_{wp}}{J_w} + \varepsilon_x \cdot |\omega - \hat{\omega}|^{0.5} \cdot \text{sgn}(\omega - \hat{\omega}) \\ \dot{\hat{F}}_x = \eta_x \cdot \text{sgn}(\omega - \hat{\omega}) \end{cases} \tag{3.19}$$

where ε_x and η_x are the yet-to-be-designed gain parameters for the ST-SMO algorithm for the longitudinal tire force.

Thus, based on the tire rotational dynamics and the identification of tire effective radius, we design an ST-SMO algorithm for each wheel separately, as shown in Eq. (3.19), to obtain a unified form for individually observing the longitudinal tire force of each wheel.

3.2.4 OBSERVATION OF TIRE LATERAL FORCE

The tire lateral force observation algorithm is based on signals from onboard sensors such as the steering wheel angle sensor and vehicle-mounted gyroscope. In the problem of tire lateral force observation, the variables that can be measured directly are vehicle steering wheel angle, yaw rate, and lateral acceleration, and the variable to be observed is the lateral tire force. Therefore, based on the vehicle dynamic model introduced in Section 2.4, we can yield the following:

$$F_{yf} = [l_r \cdot m \cdot a_y + I_z \cdot \dot{\varphi} - F_{xf}(l_f + l_r)\sin\delta]/[(l_f + l_r)\cos\delta] \tag{3.20}$$

$$F_{yr} = (l_f \cdot m \cdot a_y - I_z \cdot \dot{\varphi})/(l_f + l_r) \tag{3.21}$$

To facilitate the design of the ST-SMO algorithm for tire lateral forces, we convert the Eq. (3.20) and Eq. (3.21), which are derived from the vehicle dynamic model, into a unified state-space form as follows

$$\dot{\varphi} = B_{Fyf} \cdot u_f + D_{Fyf} \cdot F_{yf} + \Gamma_{Fyf} \tag{3.22}$$

$$\dot{\varphi} = B_{Fyr} \cdot u_f + D_{Fyr} \cdot F_{yr} + \Gamma_{Fyr} \tag{3.23}$$

where the calculation formulas of all the parameters are given by

$$\begin{cases} u_f = a_y \\ B_{Fyf} = -l_r \cdot m/I_z, D_{Fyf} = (l_f + l_r)\cos\delta/I_z, \Gamma_{Fyf} = F_{xf} \cdot (l_f + l_r)\sin\delta/I_z \\ B_{Fyr} = l_f \cdot m/I_z, D_{Fyr} = -(l_f + l_r)/I_z, \Gamma_{Fyr} = 0 \end{cases} \tag{3.24}$$

In this section, we separately design specific observers for F_{yf} and F_{yr}: for the observer of F_{yf}, its states involve φ and F_{yf}, while the states are φ and F_{yr} for the F_{yr} observer. The ST-SMO algorithms for lateral tire forces can be expressed as

$$\begin{cases} \dot{\hat{\varphi}} = D_{Fyf} \cdot \hat{F}_{yf} + \Gamma_{Fyf} + B_{Fyf} \cdot u_f + \varepsilon_{yf} \cdot |\varphi - \hat{\varphi}|^{0.5} \cdot \text{sgn}(\varphi - \hat{\varphi}) \\ \dot{\hat{F}}_{yf} = \eta_{yf} \cdot \text{sgn}(\varphi - \hat{\varphi}) \end{cases} \tag{3.25}$$

$$\begin{cases} \dot{\hat{\varphi}} = D_{Fyr} \cdot \hat{F}_{yr} + \Gamma_{Fyr} + B_{Fyr} \cdot u_f + \varepsilon_{yr} \cdot |\varphi - \hat{\varphi}|^{0.5} \cdot \text{sgn}(\varphi - \hat{\varphi}) \\ \dot{\hat{F}}_{yr} = \eta_{yr} \cdot \text{sgn}(\varphi - \hat{\varphi}) \end{cases} \tag{3.26}$$

where ε_{yf}, η_{yf}, ε_{yr}, and η_{yr} are the parameters to be designed.

Given that we have obtained the estimations of F_{yf} and F_{yr}, we can distribute lateral forces acting on each tire based on the magnitude of the vertical force on each tire, which is given by the following equation

$$\begin{cases} \hat{F}_{yfl} = \hat{F}_{yf} \cdot F_{zfl}/\left(F_{zfl} + F_{zfr}\right) \\ \hat{F}_{yfr} = \hat{F}_{yf} \cdot F_{zfr}/\left(F_{zfl} + F_{zfr}\right) \\ \hat{F}_{yrl} = \hat{F}_{yr} \cdot F_{zrl}/\left(F_{zrl} + F_{zrr}\right) \\ \hat{F}_{yrr} = \hat{F}_{yr} \cdot F_{zrr}/\left(F_{zrl} + F_{zrr}\right) \end{cases} \tag{3.27}$$

3.2.5 ESTIMATION OF TIRE SIDESLIP ANGLE

This subsection focuses on the ST-SMO algorithm for observing tire sideslip angle based on the observation results of longitudinal tire forces and lateral tire forces as discussed above. According to vehicle dynamic modeling, as introduced in Section 2.4, we can yield the calculation formulas of α_f and α_r, which are given by

$$\begin{cases} \alpha_f = \delta - \dfrac{v_{yf}}{v_x} = \delta - \dfrac{v_y + l_f \varphi}{v_x} \\ \alpha_r = -\dfrac{v_{yr}}{v_x} = -\dfrac{v_y - l_r \varphi}{v_x} \end{cases} \tag{3.28}$$

Taking the derivatives of the expressions as shown in Eq. (3.28), we can obtain:

$$\begin{cases} \dot{\alpha}_f = \dot{\delta} - \dfrac{\dot{v}_{yf}}{v_x} = \dot{\delta} - \dfrac{\dot{v}_y + l_f \dot{\varphi}}{v_x} \\ \dot{\alpha}_r = \dfrac{\dot{v}_{yr}}{v_x} = -\dfrac{\dot{v}_y - l_r \dot{\varphi}}{v_x} \end{cases} \Rightarrow \begin{cases} \dot{\varphi} = -\dfrac{v_x}{l_f}\dot{\alpha}_f - \dfrac{1}{l_f}\dot{v}_y + \dfrac{v_x}{l_f}\dot{\delta} \\ \dot{\varphi} = \dfrac{v_x}{l_r}\dot{\alpha}_r + \dfrac{1}{l_r}\dot{v}_y \end{cases} \tag{3.29}$$

By combining the equation of vehicle lateral motion, which has been introduced in Section 2.4, we can derive:

$$\dot{v}_y = -v_x \varphi + \frac{1}{m}\left[F_{yrl} + F_{yrr} + (F_{yfl} + F_{yfr})\cos\delta + (F_{xfl} + F_{xfr})\sin\delta\right] \tag{3.30}$$

Similarly, we design specific observers separately for α_f and α_r. For the α_f observer, its state variables are φ and α_f, and the state variables for the α_r observer are φ and α_r. The ST-SMO algorithms for estimating the slip angles α_f and α_r are constructed as follows:

$$\begin{cases} \dot{\hat{\varphi}} = -\dfrac{v_x}{l_f}\dot{\hat{\alpha}}_f + \dfrac{v_x}{l_f}\dot{\delta} + \dfrac{v_x}{l_f}\varphi - \dfrac{1}{l_f m}[F_{yrl} + F_{yrr} + (F_{yfl} + F_{yfr})\cos\delta \\ \quad + (F_{xfl} + F_{xfr})\sin\delta] + \varepsilon_{\alpha f} \cdot |\varphi - \hat{\varphi}|^{0.5} \cdot sgn(\varphi - \hat{\varphi}) \\ \ddot{\hat{\alpha}}_f = \eta_{\alpha f} \cdot sgn(\varphi - \hat{\varphi}) \end{cases} \tag{3.31}$$

$$\begin{cases} \dot{\hat{\varphi}} = \dfrac{v_x}{l_r}\dot{\hat{\alpha}}_r - \dfrac{v_x}{l_r}\varphi + \dfrac{1}{l_r m}[F_{yrl} + F_{yrr} + (F_{yfl} + F_{yfr})\cos\delta \\ \quad + (F_{xfl} + F_{xfr})\sin\delta] + \varepsilon_{\alpha r} \cdot |\varphi - \hat{\varphi}|^{0.5} \cdot \mathrm{sgn}(\varphi - \hat{\varphi}) \\ \ddot{\hat{\alpha}}_r = \eta_{\alpha r} \cdot \mathrm{sgn}(\varphi - \hat{\varphi}) \end{cases} \tag{3.32}$$

where $\varepsilon_{\alpha f}$, $\eta_{\alpha f}$, $\varepsilon_{\alpha r}$, and $\eta_{\alpha r}$ are the parameters to be designed for the sideslip angle ST-SMO algorithms. The integration of the observed values in the above algorithms yields the estimated values of the sideslip angles $\hat{\alpha}_f$ and $\hat{\alpha}_r$.

The tire sideslip angle ST-SMO algorithms, as presented in Eq. (3.31) and (3.32), require the longitudinal and lateral tire forces, which can be estimated by the above-designed ST-SMO algorithms for the tire-force observation. Therefore, the observers (3.31) and (3.32) designed for the slip angles α_f and α_r are solvable. Furthermore, based on the modeling of the vehicle four-wheel dynamic model, we correct the observed values $\hat{\alpha}_f$ and $\hat{\alpha}_r$ and can yield the observed values of the sideslip angle for each tire:

$$\begin{cases} \hat{\alpha}_{fl} = v_x/(v_x - \varphi T_{wb}/2) \cdot \hat{\alpha}_f + (-\varphi T_{wb}/2)/(v_x - \varphi T_{wb}/2) \cdot \delta \\ \hat{\alpha}_{fr} = v_x/(v_x + \varphi T_{wb}/2) \cdot \hat{\alpha}_f + (\varphi T_{wb}/2)/(v_x + \varphi T_{wb}/2) \cdot \delta \\ \hat{\alpha}_{rl} = v_x/(v_x - \varphi T_{wb}/2) \cdot \hat{\alpha}_r \\ \hat{\alpha}_{rr} = v_x/(v_x + \varphi T_{wb}/2) \cdot \hat{\alpha}_r \end{cases} \tag{3.33}$$

3.3 DISCRETIZATION AND STABILITY ANALYSIS

In the previous section, we have derived a unified observation system for the longitudinal tire force, lateral tire force, and tire sideslip angle of each tire based on the proposed ST-SMO algorithm and the mathematical expressions of the vehicle dynamic model. In this section, we will derive the discrete form of the ST-SMO system and analyze the stability of the algorithm.

3.3.1 DISCRETE FORM OF THE OBSERVATION SYSTEM OF TIRE DYNAMICS

When discretely implementing the ST-SMO algorithm shown in Eq. (3.8) to observe tire dynamic parameters, we iteratively compute the observations $\hat{x}_{st,1}(k+1)$ and $\hat{x}_{st,2}(k+1)$ for the $k+1$ instant via the observation error $e_{st,1}(k)$, which can be calculated based on the actual measurement $x_{st,1}(k)$ and the observation $\hat{x}_{st,1}(k)$ at the k sampling instant. The discrete ST-SMO can be described as:

$$\begin{cases} \hat{x}_{st,1}(k+1) = \hat{x}_{st,1}(k) + T_s\left[f_1(\hat{x}_{st,2}(k)) + \varepsilon_{st,1}|e_{st,1}(k)|^{0.5}\text{sgn}\left(e_{st,1}(k)\right)\right] \\ \hat{x}_{st,2}(k+1) = \hat{x}_{st,2}(k) + T_s\left[\eta_{st,1}\text{sgn}\left(e_{st,1}(k)\right) + f_2(x_{st,1}(k), \hat{x}_{st,2}(k), u_{st}(k))\right] \end{cases} \tag{3.34}$$

where T_s represents the sampling time.

Based on the general form of the discrete ST-SMO, i.e., Eq. (3.34), we can derive the discrete ST-SMO algorithms for the longitudinal tire force, lateral tire force, and tire sideslip angle.

The discrete form of the observer for longitudinal tire force is as follows:

$$\begin{cases} \hat{\omega}(k+1) = \hat{\omega}(k) + T_s \cdot [-R(k)/J_w \cdot \hat{F}_x(k) + (T_{wt}(k) \\ \qquad - T_{wp}(k))/J_w + \varepsilon_x \cdot |e_x(k)|^{0.5} \cdot \text{sgn}\left(e_x(k)\right)] \\ \hat{F}_x(k+1) = \hat{F}_x(k) + T_s \cdot \eta_x \cdot \text{sgn}\left(e_x(k)\right) \end{cases} \tag{3.35}$$

where $e_x(k) = \omega(k) - \hat{\omega}(k)$ denotes the observation error of the wheel speed.

The discrete form of the observer for lateral tire force is as follows:

$$\begin{cases} \hat{\varphi}(k+1) = \hat{\varphi}(k) + T_s \cdot [D_{Fyf}(k) \cdot \hat{F}_{yf}(k) + \Gamma_{Fyf}(k) \\ \qquad + B_{Fyf} \cdot u_f(k) + \varepsilon_{yf} \cdot |e_{yf}(k)|^{0.5} \cdot \text{sgn}\left(e_{yf}(k)\right)] \\ \hat{F}_{yf}(k+1) = \hat{F}_{yf}(k) + T_s \cdot \eta_{yf} \cdot \text{sgn}\left(e_{yf}(k)\right) \end{cases} \tag{3.36}$$

$$\begin{cases} \hat{\varphi}(k+1) = \hat{\varphi}(k) + T_s \cdot [D_{Fyr} \cdot \hat{F}_{yr}(k) + \Gamma_{Fyr} + B_{Fyr} \cdot u_f(k) \\ \qquad + \varepsilon_{yr} \cdot |e_{yr}(k)|^{0.5} \cdot \mathrm{sgn}(e_{yr}(k))] \\ \hat{F}_{yr}(k+1) = \hat{F}_{yr}(k) + T_s \cdot \eta_{yr} \cdot \mathrm{sgn}(e_{yr}(k)) \end{cases} \tag{3.37}$$

where $e_{yf}(k) = \varphi(k) - \hat{\varphi}(k)$ is the observation error of vehicle yaw rate in the observer of F_{yf}, $e_{yr}(k) = \varphi(k) - \hat{\varphi}(k)$ is the observation error of vehicle yaw rate in the observer of F_{yr}.

The discrete form of the observer for tire sideslip angle is as follows:

$$\begin{cases} \hat{\varphi}(k+1) = \hat{\varphi}(k) + T_s[v_x(k)/l_f \cdot (-\dot{\hat{\alpha}}_f(k) + \dot{\delta}(k) + \varphi(k)) \\ -(F_{yrl}(k) + F_{yrr}(k) + (F_{yfl}(k) + F_{yfr}(k))\cos\delta(k) \\ +(F_{xfl}(k) + F_{xfr}(k))\sin\delta(k))/(l_f m) + \varepsilon_{\alpha f} \cdot |e_{\alpha f}(k)|^{0.5} \cdot \mathrm{sgn}(e_{\alpha f}(k))] \\ \dot{\hat{\alpha}}_f(k+1) = \dot{\hat{\alpha}}_f(k) + T_s \cdot \eta_{\alpha f} \cdot \mathrm{sgn}(e_{\alpha f}(k)) \end{cases} \tag{3.38}$$

$$\begin{cases} \hat{\varphi}(k+1) = \hat{\varphi}(k) + T_s[v_x(k)/l_r \cdot (\dot{\hat{\alpha}}_r(k) - \varphi(k)) + (F_{yrl}(k) + F_{yrr}(k) \\ +(F_{yfl}(k) + F_{yfr}(k))\cos\delta(k) + (F_{xfl}(k) + F_{xfr}(k))\sin\delta(k))/(l_r m) \\ +\varepsilon_{\alpha r} \cdot |e_{\alpha r}(k)|^{0.5} \cdot \mathrm{sgn}(e_{\alpha r}(k))] \\ \dot{\hat{\alpha}}_r(k+1) = \dot{\hat{\alpha}}_r(k) + T_s \cdot \eta_{\alpha r} \cdot \mathrm{sgn}(e_{\alpha r}(k)) \end{cases} \tag{3.39}$$

where $e_{\alpha f}(k) = \varphi(k) - \hat{\varphi}(k)$ represents the observation error of vehicle yaw rate in the observer algorithm for α_f, and $e_{\alpha r}(k) = \varphi(k) - \hat{\varphi}(k)$ represents the observation error of vehicle yaw rate in the observer algorithm for α_r.

3.3.2 STABILITY ANALYSIS OF THE ST-SMO ALGORITHM

Some scholars have made substantial efforts to address the stability issue of super-twisting algorithms from different perspectives. The literature [80] analyzed the stability of super-twisting control by discussing the state trajectories in the phase plane. Later, the stability of high-order sliding mode algorithms was analyzed based on the homogeneity principle [79]. In addition, the Lyapunov function analysis greatly simplified the complexity of proving the stability of super-twisting algorithms [98, 118]. This book also provides a discussion about the stability of the proposed ST-SMO algorithms.

Regarding the observation states $x_{st,i}$ of the proposed ST-SMO algorithms, we define the observation error as $e_{st,i} = x_{st,i} - \hat{x}_{st,i}$, where $i = 1, 2$. According to the longitudinal tire force observer as shown in Eq. (3.35), lateral tire force observer as shown in Eqs. (3.36) and (3.37), and tire sideslip angle observer as shown in Eqs. (3.38) and (3.39), a unified formula for observation error can be derived as follows:

$$e_{st,1}(k+1) = e_{st,1}(k) + \xi_{st} \cdot T_s \cdot e_{st,2}(k) - \varepsilon_{st,1} \cdot T_s \cdot |e_{st,1}(k)|^{0.5} \cdot sgn(e_{st,1}(k)) \tag{3.40}$$

$$e_{st,2}(k+1) = e_{st,2}(k) - \eta_{st,1} \cdot T_s \cdot \mathrm{sgn}(e_{st,1}(k)) \tag{3.41}$$

The specific description of parameters in Eqs. (3.40) and (3.41), is provided in Table 3.1.

Table 3.1
Specific description of parameters in observation error equations.

Observer	$e_{st,1}$	$e_{st,2}$	ξ_{st}	$\varepsilon_{st,1}$	$\eta_{st,1}$
F_x	$\omega-\hat{\omega}$	$F_x-\hat{F}_x$	$-R/J_w$	ε_x	η_x
F_{yf}	$\varphi-\hat{\varphi}$	$F_{yf}-\hat{F}_{yf}$	$(l_f+l_r)\cos\delta/I_z$	ε_{yf}	η_{yf}
F_{yr}	$\varphi-\hat{\varphi}$	$F_{yr}-\hat{F}_{yr}$	$-(l_f+l_r)/I_z$	ε_{yr}	η_{yr}
α_f	$\varphi-\hat{\varphi}$	$\dot{\alpha}_f-\dot{\hat{\alpha}}_f$	$-v_x/l_f$	$\varepsilon_{\alpha f}$	$\eta_{\alpha f}$
α_r	$\varphi-\hat{\varphi}$	$\dot{\alpha}_r-\dot{\hat{\alpha}}_r$	v_x/l_r	$\varepsilon_{\alpha r}$	$\eta_{\alpha r}$

Then, by converting the observation error in Eqs. (3.40) and (3.41), we can yield:

$$\begin{bmatrix} e_{st,1}(k+1) \\ e_{st,2}(k+1) \end{bmatrix} = \Phi_{st} \begin{bmatrix} e_{st,1}(k) \\ e_{st,2}(k) \end{bmatrix} + \Theta_{st}\mathrm{sgn}(e_{st,1}(k)) \tag{3.42}$$

$$\Phi_{st} = \begin{bmatrix} 1 & \xi_{st}\cdot T_s \\ 0 & 1 \end{bmatrix}, \Theta_{st} = \begin{bmatrix} -\varepsilon_{st,1}\cdot T_s\cdot |e_{st,1}(k)|^{0.5} \\ -\eta_{st,1}\cdot T_s \end{bmatrix} \tag{3.43}$$

Based on Eq. (3.43), we can describe the stability of the ST-SMO algorithm as follows. Regarding the unified ST-SMO algorithm with the measurable state variable $x_{st,1}$ and the algorithm gain parameters $\varepsilon_{st,1}$ and $\eta_{st,1}$, if the inequality $\Phi_{st}^T(P_{st}+\Lambda_{st})\Phi_{st}-(1-\vartheta_{st})P_{st} < -G_{st}$ holds, where $\Lambda_{st}^{-1} = (\Lambda_{st}^{-1})^T > 0$ has a positive definite solution $P_{st} = P_{st}^T > 0$ and $\eta_{st,1} > 0$, the observation error equation shown in Eq. (3.43) converges. Thus, the stability of the observation algorithm can be ensured. The specific proof process is detailed in the literature [118], and will not be reiterated here. It is necessary to discuss the existence of solutions that satisfy the above conditions for the inequality $\Phi_{st}^T(P_{st}+\Lambda_{st})\Phi_{st}-(1-\vartheta_{st})P_{st} < -G_{st}$.

Theorem 3.1 *Given ϑ_{st} with $0<\vartheta_{st}<1$ such that the inequality (3.44) holds, then $\Phi_{st}^T(P_{st}+\Lambda_{st})\Phi_{st}-(1-\vartheta_{st})P_{st} < -G_{st}$ always has a positive definite solution $P_{st} = P_{st}^T > 0$. The solution is given by Eq. (3.45), as it can be described as a classical Lyapunov discrete equation. Alternatively, for given matrices P_{st} and G_{st}, we can always find an exact solution for ϑ_{st}.*

$$\left| \frac{\lambda_{\min}\{\Phi_{st}\}}{\sqrt{1-\vartheta_{st}}} \right| < 1 \tag{3.44}$$

$$P_{st} = \sum_{k=0}^{\infty} \left(\tilde{\Phi}_{st}^T\right)^k G_{st,1}\left(\tilde{\Phi}_{st}\right)^k \tag{3.45}$$

Proof *Translating the inequality $\Phi_{st}^T(P_{st}+\Lambda_{st})\Phi_{st}-(1-\vartheta_{st})P_{st} < -G_{st}$ into $\Phi_{st}^T P_{st}\Phi_{st}-(1-\vartheta_{st})P_{st} < -G_{st}-\Phi_{st}^T\Lambda_{st}\Phi_{st}$, we can derive the following:*

$$\Phi_{st}^T P_{st}\Phi_{st}-(1-\vartheta_{st})P_{st} \leq -\tilde{G}_{st} \tag{3.46}$$

where $\tilde{G}_{st} = G_{st} + \Phi_{st}^T \Lambda_{st} \Phi_{st}$.

Furthermore, given that $\tilde{\vartheta}_{st} = 1 - \vartheta_{st} > 0$, *we can divide both sides of the above equation by* $\tilde{\vartheta}_{st}$ *and yield:*

$$\tilde{\Phi}_{st}^T P_{st} \tilde{\Phi}_{st} - P_{st} = -G_{st,1} \tag{3.47}$$

where $\tilde{\Phi}_{st} = \sqrt{\tilde{\vartheta}_{st}^{-1}} \Phi_{st}$ *and* $G_{st,1} = \tilde{\vartheta}_{st}^{-1} \tilde{G}_{st}$.

Clearly, Eq. (3.47) represents a discrete-time Lyapunov equation, and its explicit solution is given by Eq. (3.45).

3.4 SIMULATION ANALYSIS OF TIRE DYNAMICS ESTIMATION

This section provides a detailed simulation and analysis of tire dynamics estimation. A co-simulation platform is built, and it involves the required vehicle dynamic model, a variety of simulation scenarios, and necessary signal interfaces.

3.4.1 DESIGN OF THE SIMULATION SCENARIO

This chapter investigates the observer system for the longitudinal tire force, lateral tire force, vertical tire force, and tire sideslip angle of four wheels. This section comprehensively considers the typical scenes for verifying each estimator. The typical situation for verifying the longitudinal tire force estimation involves vehicle braking and driving conditions. The typical conditions for verifying the vertical tire force, lateral tire force, and tire sideslip angle are continuous steering maneuvers. Therefore, this section designs the composite-condition scenario, as shown in Fig. 3.7, to validate the proposed observer system in this chapter. In the initial 14 s of the composite-condition simulation test, the vehicle operates an acceleration maneuver with a throttle

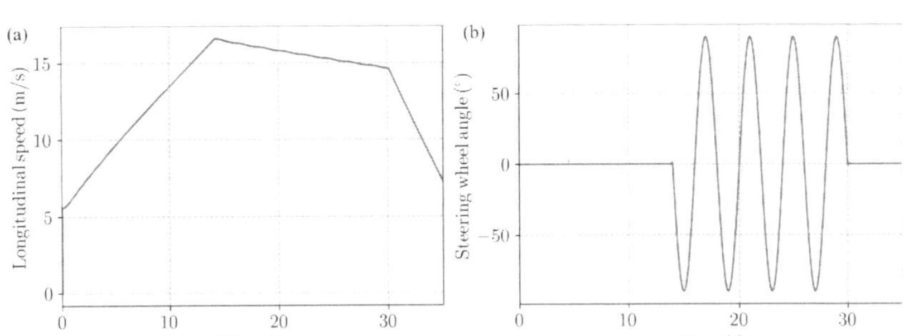

Figure 3.7 Vehicle speed and steering wheel angle in the simulation of the composite-condition scenario.

opening of 0.2, accelerating the vehicle from an initial speed of 5.5 m/s to approximately 17 m/s. Following this, the vehicle operates a sinusoidal-steering maneuver with an amplitude of 90° and a steering frequency of 4 Hz. 30 s after the simulation started, the vehicle performs a braking maneuver with a master cylinder braking pressure of 1 MPa. In addition, this test is performed on a road with a friction coefficient of 0.3.

Fig. 3.8 shows vehicle lateral acceleration and yaw rate during the composite-condition simulation. As the sinusoidal-steering wheel angle input varies, the vehicle lateral acceleration and yaw rate change correspondingly, exhibiting similar trends among the three variables. As shown in Figs. 3.7 and 3.8, vehicle yaw rate reaches its maximum value at 29 s, with a peak value of 12.3°/s, and the maximum lateral acceleration can reach $0.29g$, exceeding $0.75\mu g$. This indicates that the vehicle and tire dynamics have entered the nonlinear region, where the vehicle is susceptible to sideslip and instability.

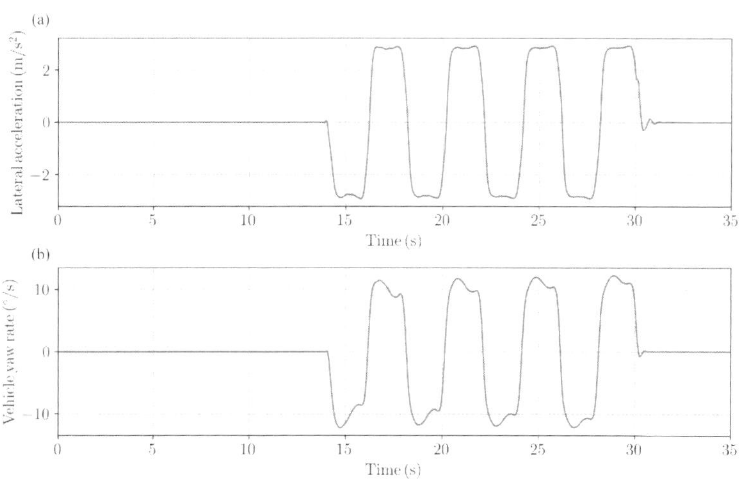

Figure 3.8 Vehicle lateral acceleration and yaw rate in the composite-condition scenario.

3.4.2 ANALYSIS OF SIMULATION RESULTS

This subsection provides the estimation results and error analysis of the longitudinal tire force, lateral tire force, vertical tire force, and tire sideslip angle in the composite-condition simulation.

Fig. 3.9 demonstrates the estimation results of the longitudinal tire force, which compares the traditional PID observer, the ST-SMO based on the identification of effective tire radius, and the actual longitudinal tire force. During the process of acceleration, which lasts from 0 s to 15 s, the longitudinal tire forces of the left front wheel and right front wheel are approximately 650 N,

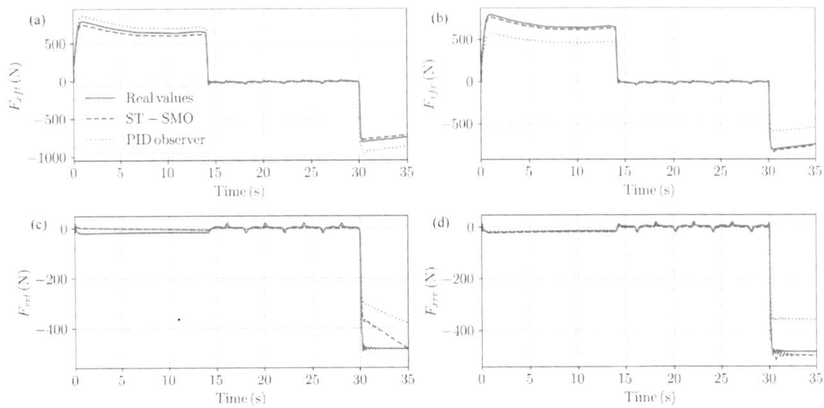

Figure 3.9 Simulation results of longitudinal tire forces in the composite-condition scenario.

Table 3.2

Comparison and analysis of longitudinal tire force estimation errors in the composite-condition scenario.

Estimation algorithm	Metric	F_{xfl}	F_{xfr}	F_{xrl}	F_{xrr}
PID observer	Peak value (N)	811.3	809.7	493.2	489.1
	RMSE (N)	66.23	144.6	56.66	53.59
	Error ratio (%)	8.16	17.86	11.49	10.96
ST-SMO	Peak value (N)	811.3	809.7	493.2	489.1
	RMSE (N)	29.24	15.17	27.52	7.27
	Error ratio (%)	3.60	1.87	5.58	1.49

while the longitudinal tire forces of the left rear wheel and right rear wheel are about 0 N. The longitudinal tire forces of four wheels fluctuate slightly around 0 N when the vehicle performs the sinusoidal-steering maneuver. In the braking phase, the longitudinal tire forces of the front wheels are about −750 N, and the longitudinal tire forces of the rear wheels are about −480 N. From Fig. 3.9, it is evident that the longitudinal tire forces estimated by the method proposed in this chapter closely match the actual tire force curves. In contrast, the observation curves from the PID observer show significant deviations from the actual values.

The detailed error analysis is presented in Table 3.2. The root-mean-square-errors (RMSEs) of the PID observer for the longitudinal tire forces of the left front wheel, right front wheel, left rear wheel, and right rear wheel are 66.23 N, 144.6 N, 56.66 N, and 53.59 N, respectively, with the error percentage of around 10%. In contrast, the ST-SMO algorithm for the observation of

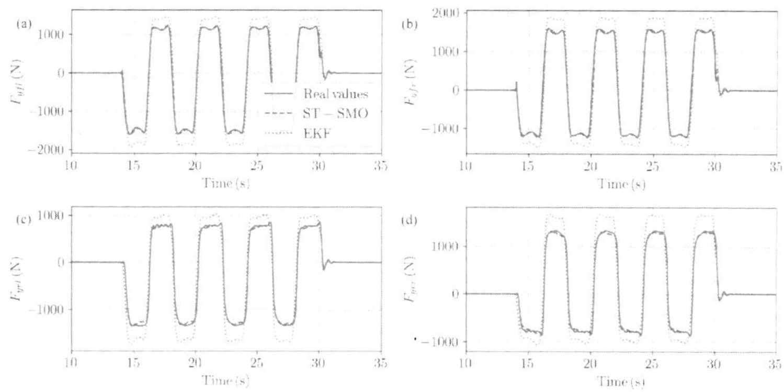

Figure 3.10 Simulation results of lateral tire forces in the composite-condition scenario.

longitudinal tire forces has RMSEs of 29.24 N, 15.17 N, 28.25 N, and 8.04 N for respective wheels, all smaller than the PID observer's error values, with error percentages of 3.60%, 1.87%, 5.73%, and 1.64%, respectively. This suggests the proposed ST-SMO algorithm, which utilizes the identification of effective tire radius, has significant advantages over the PID observer.

Fig. 3.10 shows the observation results of lateral tire forces in the composite-condition scenario. During the sinusoidal steering, the lateral tire forces of four tires fluctuate in response to the steering wheel angle, exhibiting a sinusoidal pattern. When the steering wheel angle is $0°$, the lateral tire forces of the four wheels are approximately 0 N. The maximum values of lateral tire forces of the two front wheels are about 1600 N, and the maximum values of the two rear wheels are about 1300 N. Combining with Fig. 3.8, it can be seen that when vehicle lateral acceleration is within a small range (less than $0.5\mu g$), the estimated lateral tire forces for each wheel, as derived from both the EKF and the ST-SMO algorithms, closely match the actual curves. This indicates both methods achieve high observation accuracy in these conditions. When vehicle lateral acceleration exceeds $0.75\mu g$, the lateral tire forces for all four wheels reach their peak values. In these conditions, the observation results from the EKF show significant deviations from the actual values. However, the estimation curves of the proposed observer still closely align with the actual curves, indicating that the ST-SMO algorithm for lateral tire forces has higher estimation accuracy than the EKF observer in the nonlinear region of vehicle dynamics.

Details of estimated errors of lateral tire forces are illustrated in Table 3.3. It can be seen that the EKF algorithm has relatively large errors in lateral tire force observations for each wheel, with the RMSEs exceeding 200 N and error percentages ranging from 10% to 15%. The ST-SMO method shows considerable performance superiority over the EKF: the RMSE values of the lateral tire forces for the left front wheel, right front wheel, left rear wheel,

Table 3.3

Comparison and analysis of lateral tire force estimation errors in the composite-condition scenario.

Estimation algorithm	Metric	F_{yfl}	F_{yfr}	F_{yrl}	F_{yrr}
EKF	Peak value (N)	1602	1594	1338	1341
	RMSE (N)	222.4	228.6	209.8	208.1
	Error ratio (%)	13.89	14.34	15.68	15.52
ST-SMO	Peak value (N)	1602	1594	1338	1341
	RMSE (N)	66.4	72.0	26.6	25.2
	Error ratio (%)	4.15	4.52	1.99	1.88

and right rear wheel are 66.4 N, 72.0 N, 26.6 N, and 25.2 N, respectively, with corresponding error percentages of 4.15%, 4.52%, 1.99%, and 1.88%. It suggests that the observation accuracy of the proposed method is improved by about 10% compared to the EKF.

The estimation results of tire sideslip angle by various observers are demonstrated in Fig. 3.11. Similar to the lateral tire force curves of each wheel, the tire sideslip angles change with the steering wheel angle, showing a sinusoidal variation trend, when the vehicle performs the sinusoidal-steering maneuver. The peak values of the front-wheel tire sideslip angles are about 5.5°, while the peak values of the rear-wheel tire sideslip angles are about 2.5°, which is relatively smaller. Similar to the observation results of lateral tire forces, both the EKF and the ST-SMO algorithms have high accuracy in observing tire sideslip angles in the linear region (when the tire sideslip angles are small, and the vehicle lateral acceleration is less than $0.5\mu g$). However, when the

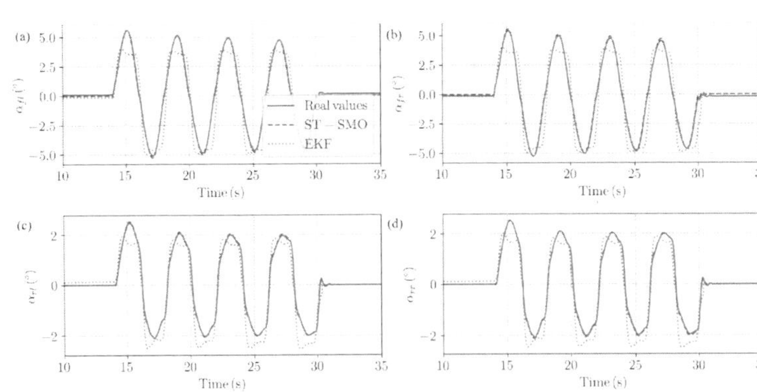

Figure 3.11 Simulation results of tire sideslip angles in the composite-condition scenario.

Table 3.4
Comparison and analysis of tire sideslip angle estimation errors in the composite-condition scenario.

Estimation algorithm	Metric	α_{fl}	α_{fr}	α_{rl}	α_{rr}
	Peak value (°)	5.57	5.41	2.47	2.51
EKF	RMSE (°)	0.96	0.89	0.36	0.36
	Error ratio (%)	17.21	16.42	14.57	14.25
	Peak value (°)	5.57	5.41	2.47	2.51
ST-SMO	RMSE (°)	0.13	0.14	0.04	0.04
	Error ratio (%)	2.39	2.55	1.62	1.59

tire sideslip angles of each wheel are large, the observation error of the EKF algorithm becomes significantly larger compared to the ST-SMO algorithm.

Table 3.4 presents the comparative analysis of errors in tire sideslip angle observations in the composite-condition scenario. In general, the ST-SMO algorithm introduced in this chapter demonstrates approximately 10% higher accuracy in observing tire sideslip angles compared to the EKF. Concretely, the RMSE values of the left front wheel, right front wheel, left rear wheel, and right rear wheel, achieved by the EKF, are 0.96°, 0.89°, 0.36°, and 0.36°, respectively, with corresponding error percentages of 17.21%, 16.42%, 14.57%, and 14.25%. In contrast, the RMSE values achieved by the proposed ST-SMO algorithm are 0.13°, 0.14°, 0.04°, and 0.04°, respectively, with corresponding error percentages of 2.39%, 2.55%, 1.62%, and 1.59%. It is evident that the ST-SMO algorithm proposed in this chapter outperforms the EKF algorithm in observing tire sideslip angles.

Fig. 3.12 and Table 3.5 present the estimation results of vehicle vertical tire force. When the vehicle performs straight-line driving, the vertical tire forces on the symmetrical left and right wheels are approximately equal. The vehicle load shifts to the rear axle during the process of acceleration such that the front-wheel vertical tire forces are approximately 4500 N and the rear-wheel vertical tire forces are approximately 3700 N. Conversely, when the vehicle performs the braking maneuver, the vehicle load shifts to the front axle, which increases the front-wheel vertical tire forces to about 4850 N and decreases rear-wheel vertical tire forces to about 3350 N. During the sinusoidal-steering maneuver, the vertical force varies in a sinusoidal pattern corresponding to the steering wheel angle. As shown in Table 3.5, the tire vertical force calculation method, which considers the impact of longitudinal and lateral accelerations on the vertical load of each wheel, demonstrates high accuracy. The RMSE values in the composite-condition scenario are within 150 N, with error percentages below 4%.

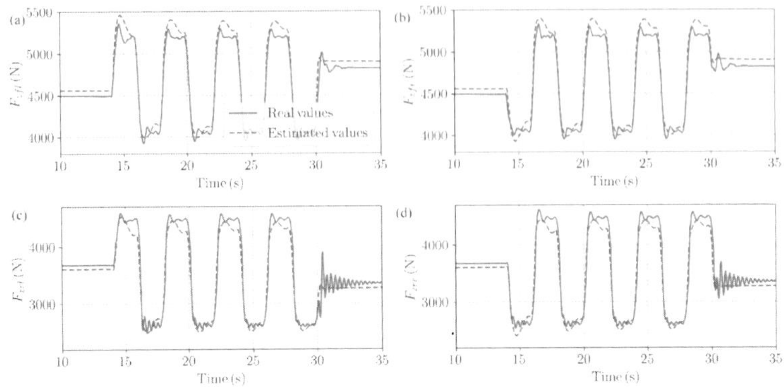

Figure 3.12 Simulation results of vertical tire forces in the composite-condition scenario.

Table 3.5
Error analysis of vertical tire force estimation in the composite-condition scenario.

Metric	F_{zfl}	F_{zfr}	F_{zrl}	F_{zrr}
Peak value (N)	5458	5399	4530	4449
RMSE (N)	139.2	138.5	153.3	150.1
Error ratio (%)	2.55	2.57	3.38	3.37

3.5 CONSTRUCTION OF EXPERIMENTAL PLATFORM

The full-condition dynamic states observation platform is constructed to validate the proposed estimation system in this chapter. This section provides a detailed introduction to the construction of the full-condition dynamic states observation platform. The experimental platform is built on a sedan car as shown in Fig. 3.13, and WFTs, dual-axis optical sensors, and other onboard sensors are utilized to measure corresponding signals.

3.5.1 SPECIFICATION OF THE EXPERIMENTAL SEDAN CAR

The utilized sedan car, as shown in Fig. 3.13, is employed to construct the full-condition platform. The main parameters of the sedan car are listed in Table 3.6.

The vehicle's onboard sensors include the steering wheel angle sensor, gyroscope, wheel speed sensors, etc. Apart from the WFTs and dual-axis optical sensors, all other required test signals generated by the EMS, ESP, and other critical chassis components are transmitted via the CAN. The key signals and their corresponding CAN protocol are shown in Table 3.7.

Figure 3.13 Photo of the employed sedan car.

Table 3.6
Main parameters of the employed sedan car.

Parameter	Unit	Value
Length × Width × Height	mm	4880 × 1800 × 1450
Wheelbase	mm	2790
Track width	mm	1560
Vehicle mass	kg	1654
Distance from vehicle mass center to front axle	mm	1200
Distance from vehicle mass center to rear axle	mm	1590
Height of vehicle mass center	mm	510
Final drive ratio	−−	3.311
Tire specification	−−	$205/55R16$

3.5.2 WHEEL FORCE TRANSDUCER

The WFT is a key transducer used for vehicle performance testing and analysis of vehicle chassis dynamics. It can measure the dynamic variations of tire forces in various test conditions including tire load, longitudinal tire force, lateral tire force, and braking/driving torques. In this book, the WFT utilized in the experimental platform is the RoaDyn Wheel Force Sensor S635 produced by the Swiss Kistler Group, as shown in Fig. 3.14. This WFT requires a customized hub adapter, which matches the employed sedan car and is installed on the vehicle's axle. Vehicle tires are mounted on the custom-modified wheel rim adapter, and the force sensor is installed between the two adapters. When the car performs maneuvers, the forces acting on the tire are transmitted to the entire vehicle and also to the WFT force sensor. Tire forces are accurately measured through sensor acquisition, integrated signal

Table 3.7
Vehicle CAN protocol of the employed sedan car.

Vehicle signal	Message ID	Message cycle	Transmission component
Engine speed	$0x138$	10 ms	EMS
Engine torque	$0x138$	10 ms	EMS
Automatic transmission gear position	$0x188$	10 ms	TCU
Wheel speed	$0x141$	20 ms	ESP
Reference speed	$0x241$	20 ms	ESP
Steering wheel angle	$0x80$	10 ms	SAS
Steering wheel angular velocity	$0x80$	10 ms	SAS
Vehicle yaw rate	$0x2F9$	10 ms	Gyroscope
Vehicle lateral acceleration	$0x2F9$	10 ms	Gyroscope

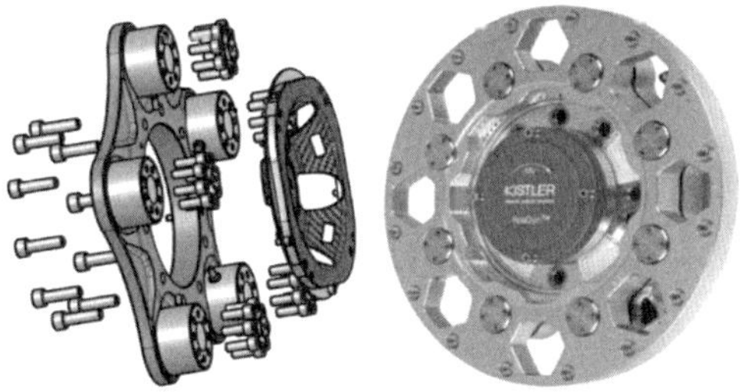

Figure 3.14 Explosive view and physical view of the RoaDyn Wheel Force Sensor S635.

amplification, and signal calculation. The five independent force measurement units in the WFT generate corresponding deformation measurements when a car conducts driving experiments. The measurement results of the five force units are decoupled to obtain the forces and moments acting on the wheels in three directions.

After calibration and testing, the RoaDyn Wheel Force Sensor S635 can perform sufficient measurement range, high measurement accuracy, and adequate sensitivity for tire forces in various directions, as shown in Table 3.8.

3.5.3 DUAL-AXIS OPTICAL SENSOR SFII

The Correvit SFII sensor, produced by the Swiss Kistler Group, is used for measuring longitudinal and lateral dynamic parameters in experiments of high-speed driving and extreme-condition tests. The picture of the Correvit SFII sensor is shown in Fig. 3.15. Leveraging long-life, anti-vibration

Table 3.8
Measurement specifications of the RoaDyn Wheel Force Sensor S635.

Tire force	Range	Sensitivity	Linearity (±%FSO)	Uncertainty (%)
F_x	$-35 \sim 35$ kN	1 N/N	0.02	0.08
F_y	$-20 \sim 20$ kN	1 N/N	0.06	0.20
F_z	$-35 \sim 35$ kN	1 N/N	0.02	0.08
M_x	$-5 \sim 5$ kN · m	1 N · m/N · m	0.09	0.30
M_y	$-5 \sim 5$ kN · m	1 N · m/N · m	0.01	0.05
M_z	$-5 \sim 5$ kN · m	1 N · m/N · m	0.11	0.30

infrared LED optical sensors, the Correvit SFII can accurately measure the traveled distance, longitudinal speed, lateral speed, and steering angle, and it can operate measurement at a calibration speed up to 250 km/h. The sensor is compact, lightweight, and equipped with protective optical glass lenses to safeguard the optical components and light sources. This ensures ease of installation and reliable operation in various conditions. The SFII sensor has 4 analog and 4 digital output channels, facilitating high-speed data transmission through CAN bus or RS-232C. Its performance specifications for measuring signals are detailed in Table 3.9.

3.5.4 CONSTRUCTION PROCEDURE OF THE FULL-CONDITION PLATFORM

To construct the experimental platform, we use four sets of WFTs and their custom adapters, a set of WFT installation brackets, a calibration remote controller, two sets of the Correvit SFII sensors and their signal modulation boxes, a set of signal conditioners, data acquisition and processing equipment, and a set of tools.

The main steps for constructing the comprehensive vehicle dynamics measurement platform are illustrated in Figs. 3.16 to 3.19. First, using preset

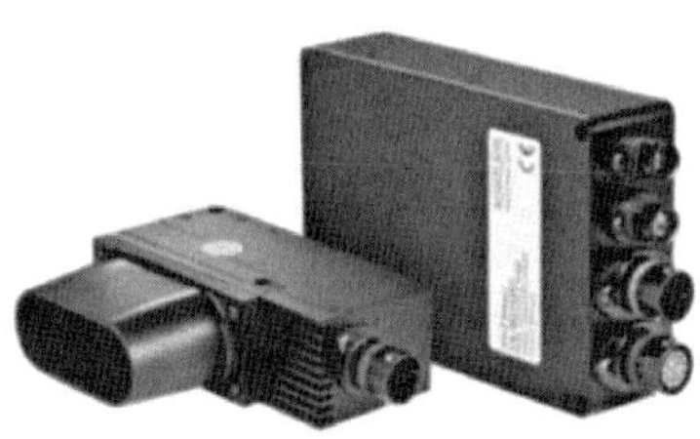

Figure 3.15 Illustration of the Correvit SFII sensor.

Table 3.9
Technical specifications of the Correvit SFII sensor.

Technical indicator	Unit	Value
Speed range	km/h	$0.3 \sim 250$
Measurement accuracy	%FSO	$\leqslant \pm 0.5$
Steering angle range	$^{\circ}$	± 40
Steering angle resolution	$^{\circ}$	$\leqslant \pm 0.1$
Measurement frequency	Hz	250
Operating range	mm	180 ± 50
Interface	--	CAN, RS-232C

torque wrenches, we install all force units and other bolts with a constant torque, and the wheel force sensor modules are installed and mounted on the tire adapters (Fig. 3.16). Next, we lift the employed sedan car with a hoist and then remove the original tires of the vehicle. The original tires are replaced with adapters equipped with wheel force sensors (Fig. 3.17). Then, we install two sets of Correvit SFII sensors and other relevant sensors. After all sensors are installed, the installation brackets are mounted to fix the sensor harnesses, and the respective signal modulation boxes and signal conditioners are placed in the vehicle. In addition, we connect correctly all sensor harnesses and power harnesses (Fig. 3.18). Furthermore, we have to perform the angle and force calibrations for all WFTs. We use a spirit level to calibrate the angle of all

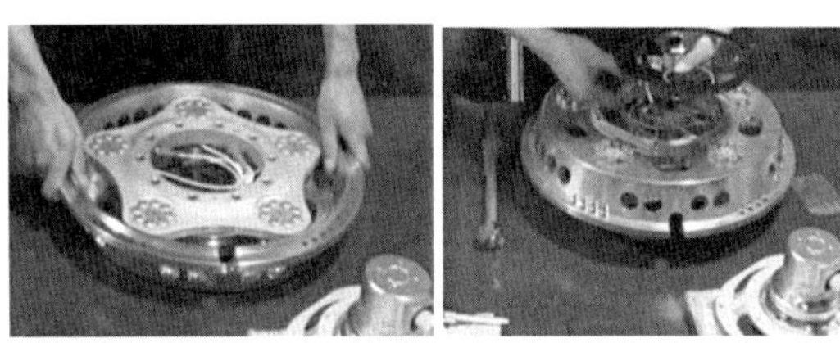

Figure 3.16 Assembly of the wheel force sensor modules.

Figure 3.17 Installation of tire adapters and Correvit SFII sensors on the sedan car.

Figure 3.18 Connection of sensor harnesses and installation of brackets.

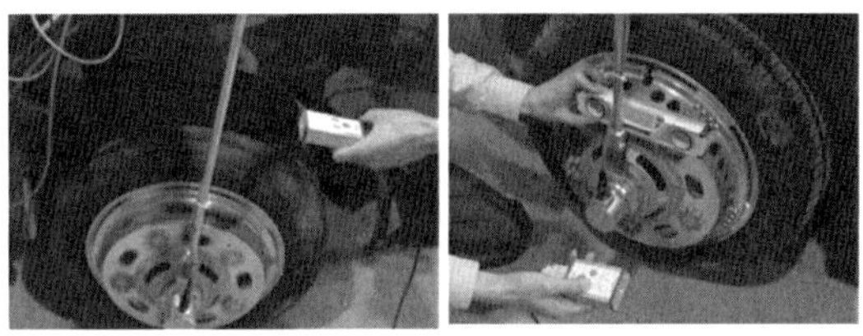

Figure 3.19 Angle and force calibration of wheel force sensors.

WFTs. Then, we conduct the force calibration procedure when the vehicle is suspended by the hoist and the wheels are spinning freely (Fig. 3.19).

Thus, the comprehensive vehicle dynamics measurement platform can be constructed as shown in Fig. 3.20. The WFTs are used to measure tire forces and torques in three directions. The Correvit SFII sensors installed on the vehicle's front axle measure the longitudinal and lateral velocities as well as the steering angle of the front two wheels, from which the tire sideslip angles of the two wheels can be calculated. Key signals such as lateral acceleration, yaw rate, wheel speeds, and steering angle are measured by vehicle onboard sensors. Signal processing is performed by data acquisition and processing equipment, and data acquisition and recording are carried out by the acquisition devices of CAN signals.

3.6 REAL-CAR EXPERIMENTAL TESTS OF TIRE DYNAMICS ESTIMATION

3.6.1 DESIGN OF THE EXPERIMENTAL SCENARIO

We employ the constructed experimental platform as introduced in the previous section to verify the tire dynamic estimation method proposed in this chapter. The experiment is conducted on a closed asphalt road with a road surface adhesion coefficient of about 0.7, as shown in Fig. 3.21. After the real-car test starts, the initial 20 s are the debugging phase for the experimental driver and data collection personnel. Then, the experimental driver first accelerates strongly by pressing the throttle pedal, then releases the throttle pedal and performs a serpentine steering maneuver. Subsequently, the driver presses

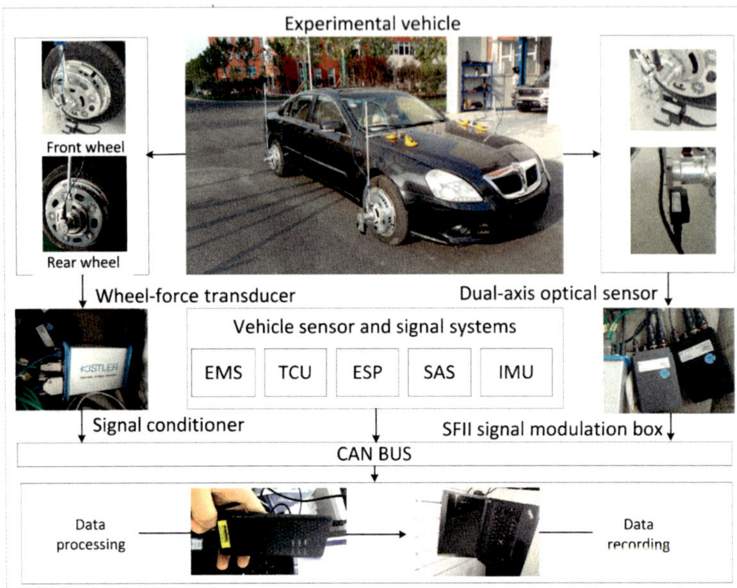

Figure 3.20 Demonstration of the comprehensive vehicle dynamics measurement platform.

Figure 3.21 Demonstration of the pre-test site and asphalt test road.

the brake pedal to perform strong deceleration. The steering wheel angle and vehicle speed during the experimental test are shown in Fig. 3.22. Concretely, we perform acceleration maneuvers with the full throttle open for about 18s to reach a speed of around 60 km/h. Then, a serpentine steering operation is performed with a steering cycle of about 5 s and a steering amplitude of around 100°. An emergency braking maneuver is performed after the steering operation to stop the test car.

Regarding this experiment, vehicle lateral acceleration and yaw rate are shown in Fig. 3.23. During the process of straight-line acceleration and

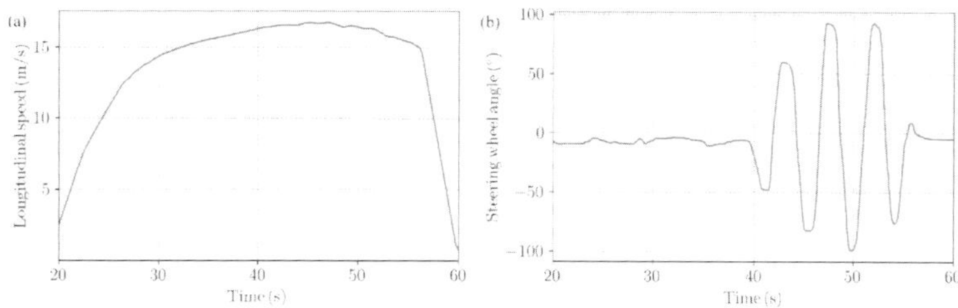

Figure 3.22 Vehicle speed and steering wheel angle in the real-car experiment.

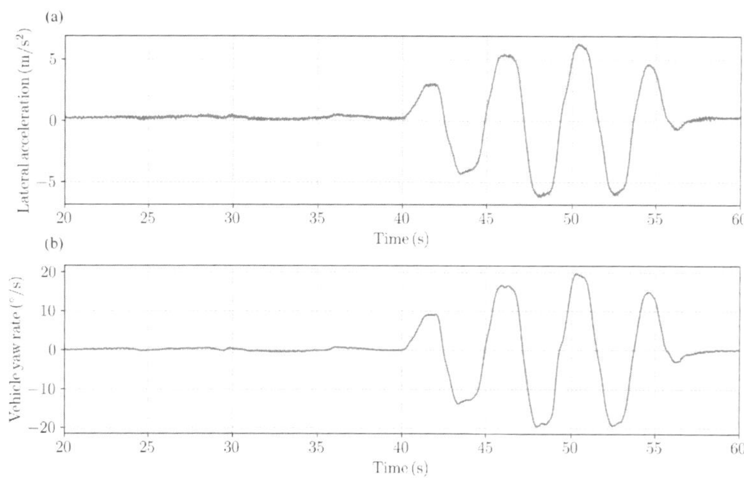

Figure 3.23 Vehicle lateral acceleration and yaw rate in this real-car experiment.

deceleration, vehicle lateral acceleration and yaw rate remain approximately zero. During serpentine steering maneuvers, vehicle lateral acceleration and yaw rate exhibit the same trend as the steering wheel angle. Vehicle lateral acceleration reaches its peak value of $0.64g$ at about 50 s, which is greater than $0.75\mu g$ (calculated based on the road surface adhesion coefficient of 0.7), and the peak value of vehicle yaw rate is about $20°/s$.

3.6.2 ANALYSIS OF EXPERIMENTAL RESULTS

Here we will analyze the experimental results of the estimation of the longitudinal tire force, lateral tire force, tire sideslip angle, and vertical tire force.

Fig. 3.24 and Table 3.10 provide detailed experimental results of the longitudinal tire force observation. As shown in Fig. 3.24, the estimated values of the proposed algorithm and the PID observer can track the actual values measured by the WFTs in this experiment. From 20 s to 30 s, during the acceleration phase, the longitudinal tire forces on the left and right front wheels are positive, gradually decreasing from about 2000 N to 0 N, while the forces on the rear wheels remain close to 0 N. Subsequently, until 57 s, the longitudinal tire forces of all four wheels remain essentially at 0 N. During the braking deceleration phase, the longitudinal tire forces of all four wheels become negative, with peak values of about −3800 N on the front wheels and −1300 N on the rear wheels. Moreover, the estimation accuracy of the proposed algorithm is higher. It can be seen from Fig. 3.24 that the estimation curve of the proposed algorithm closely aligns with the measurement curve from WFTs. In contrast, the PID observer shows significant deviations, particularly during the strong deceleration phase. Table 3.10 shows that the proposed observer for longitudinal tire force has an accuracy improvement of more than 5% compared to the PID observer algorithm.

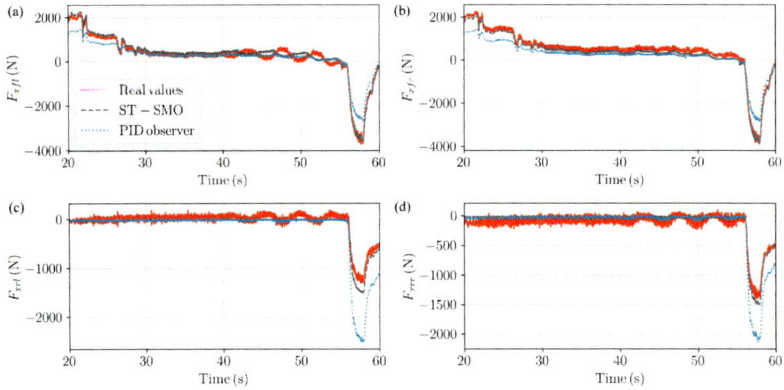

Figure 3.24 Experimental results of longitudinal tire force estimation in this real-car experiment.

Table 3.10

Error comparison and analysis of longitudinal tire force estimation in the real-car experiment.

Estimation algorithm	Metric	F_{xfl}	F_{xfr}	F_{xrl}	F_{xrr}
PID observer	Peak value (N)	3713	3884	1305	1397
	RMSE (N)	298.5	387.3	302.5	184.4
	Error ratio (%)	8.04	9.97	23.18	13.20
ST-SMO	Peak value (N)	3713	3884	1305	1397
	RMSE (N)	156.0	149.1	83.15	76.75
	Error ratio (%)	4.20	3.84	6.37	5.49

Fig. 3.25 and Table 3.11 provide detailed results of the lateral tire force observation in the real-car experiment. It can be seen from Fig. 3.25 that both the proposed ST-SMO and the EKF algorithms can track the actual tire force curves measured by the WFTs. During the process of performing steering maneuvers from 40 s to 55 s, the lateral tire force for each wheel exhibits a sinusoidal pattern. Similar to the simulation results, when the lateral tire forces are relatively large, the curves of the EKF algorithm significantly deviate from the measured data obtained by the WFTs. Detailed analysis is shown in Table 3.11. The RMSE values achieved by the EKF algorithm are all around 400 N, and the error percentages for the left front, right front, left rear, and right rear wheels are 8.30%, 8.93%, 13.92%, and 13.88%, respectively. Regarding the proposed ST-SMO, its RMSE values of the estimated lateral tire forces are 217.4 N, 226.3 N, 176.0 N, and 165.8 N, respectively, and the corresponding error percentages are 4.83%, 5.21%, 5.83%, and 5.87%, respectively. Overall, the ST-SMO method improves the observation accuracy of lateral tire force by an average of 5.8% compared to the EKF observer.

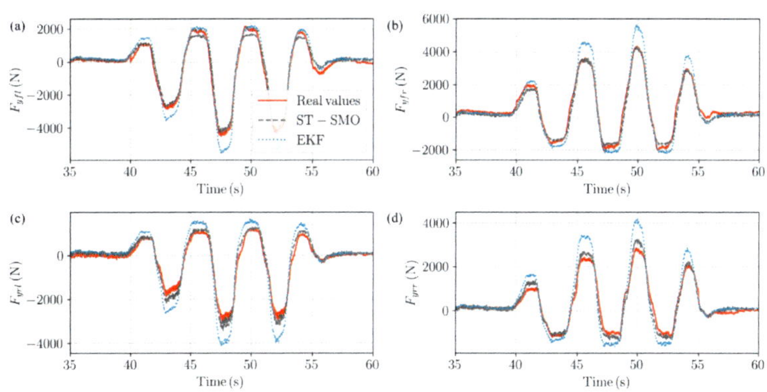

Figure 3.25 Experimental results of lateral tire forces in the real-car experiment.

Table 3.11

Error comparison and analysis of lateral tire force estimation in the real-car experiment.

Estimation algorithm	Metric	F_{yfl}	F_{yfr}	F_{yrl}	F_{yrr}
	Peak value (N)	4504	4345	3017	2867
EKF	RMSE (N)	373.9	387.9	419.9	397.9
	Error ratio (%)	8.30	8.93	13.92	13.88
	Peak value (N)	4504	4345	3017	2867
ST-SMO	RMSE (N)	215.06	223.07	173.81	165.56
	Error ratio (%)	4.77	5.13	5.76	5.77

Fig. 3.26 and Table 3.12 present detailed results of the tire sideslip angle observation. We equip the experimental car with two sets of dual-axis optical SFII sensors to directly measure the tire sideslip angle values of the two front wheels. Due to the limitation of SFII sensors, we can not obtain the actual tire sideslip angles of the rear wheels. The raw signals measured by the SFII sensors based on optical principles exhibit significant noise (denoted by the yellow dashed line in Fig. 3.26), so we conduct a simple mean filtering process and obtain the filtered data (the blue solid line in Fig. 3.26). It can be seen from Fig. 3.26 that the estimated values of both the proposed ST-SMO algorithm and the EKF estimator can track the measured values from the SFII sensors. in the real-car experiment. When the vehicle performs the steering maneuvers from 40 s to 55 s, the tire sideslip angle of each wheel exhibits a sinusoidal-like curve, otherwise, it is approximately 0°. Similar to the simulation results, both EKF and ST-SMO algorithms exhibit high accuracy in estimating tire sideslip angles within the tire linear regions (where tire sideslip angles are small), while the EKF algorithm exhibits larger errors compared to the ST-SMO method when the tire sideslip angles are large.

Similar to the simulation results, when the values of tire sideslip angles are large, which suggests the tire dynamics have entered the nonlinear region, the deviation between the EKF algorithm and the measured data is significant. Due to the limitations of experimental equipment, the experimental platform only mounts the SFII sensors on the front wheels. Direct measurements of the tire sideslip angles of the rear wheels are not conducted.

Table 3.12 presents the observed results of tire sideslip angles of the left front wheel and right front wheel. When using the EKF algorithm to observe the tire sideslip angles, the RMSE values for the left front wheel and right front wheel are 0.866° and 0.853°, respectively, with corresponding error percentages of 29.75% and 25.95%. When using the proposed ST-SMO algorithm, the RMSE values of the observations for the left front wheel and right front wheel are 0.283° and 0.275°, respectively, with corresponding error percentages of 9.73% and 9.45%. The observation errors in tire sideslip angles in this real-car experiment are slightly larger than the simulation results, which is caused by the measurement noises of the SFII sensors. Overall, the ST-SMO method improves the observation accuracy of tire sideslip angle by over 15% compared to the EKF observer in the real-car experiment.

Fig. 3.27 and Table 3.13 provide detailed observation results of the vertical tire forces in the real-car experiment. The vertical tire forces of symmetrically positioned left and right wheels are generally similar when the vehicle performs straight-line driving. During steering maneuvers, the vertical tire forces exhibit sinusoidal-like variations corresponding to changes in the steering wheel angle. As shown in Table 3.13, the accuracy of the vertical tire force calculation method is high. In this real-car experiment, the RMSE values of the observed vertical tire forces for the left front wheel, right front wheel, left rear wheel, and right rear wheel are 633.6 N, 628.0 N, 623.9 N, and

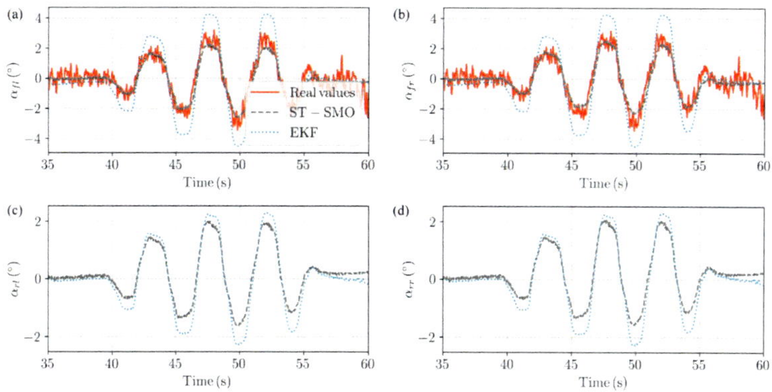

Figure 3.26 Experimental results of tire sideslip angle estimation in the real-car experiment.

Table 3.12

Error comparison and analysis of tire sideslip angle estimation in the real-car experiment.

Estimation algorithm	Metric	α_{fl}	α_{fr}
EKF	Peak value (°)	2.91	2.91
	RMSE (°)	0.866	0.853
	Error ratio (%)	29.75	25.95
ST-SMO	Peak value (°)	2.91	2.91
	RMSE (°)	0.283	0.275
	Error ratio (%)	9.73	9.45

625.8 N, respectively, with corresponding error percentages of 8.96%, 9.14%, 11.81%, and 12.16%. Although the estimated results of vertical tire forces in the real-car experiment meet the requirements, there is a slight decrease in accuracy compared to the simulation results. We can see from Fig. 3.27 that the main reason for the increased calculation errors in vertical tire forces is the discrepancy between the calculated vertical tire forces and the WFTs' measurements, which occurs around 56 s. This phenomenon is attributed to the *nodding effect* of the experimental sedan car when the vehicle performs strong emergency braking maneuvers, leading to a transfer of loads between the front and rear axles. However, during this period, vehicle longitudinal deceleration is approximately 0, and since mass-produced cars do not have inertial measurement units, the *nodding effect* cannot be quantified. Considering the practicality and cost considerations associated with mass-produced vehicles, this book does not address the influence of the *nodding effect*, but this aspect could be further investigated in future research. Nevertheless, the accuracy of the proposed method for calculating vertical tire forces meets the requirements.

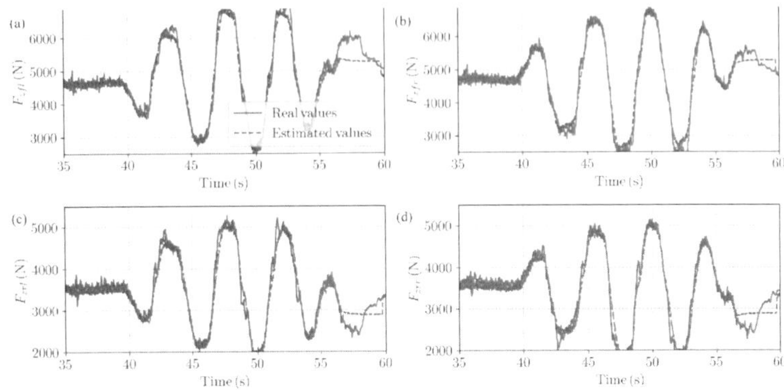

Figure 3.27 Experimental results of vertical tire forces in the real-car experiment.

Table 3.13

Error comparison and analysis of vertical tire force estimation in the real-car experiment.

Metric	F_{zfl}	F_{zfr}	F_{zrl}	F_{zrr}
Peak value (N)	7071	6869	5281	5146
RMSE (N)	633.6	628.0	623.9	625.8
Error ratio (%)	8.96	9.14	11.81	12.16

3.7 CONCLUSION

This chapter investigates the unified observation system for tire dynamic parameters (including longitudinal tire force, lateral tire force, vertical tire force, and tire sideslip angle), and validates the performance of the proposed observation system through simulation analysis and real-car experiments. First, a vertical tire force calculation method based on vehicle longitudinal and lateral accelerations is proposed. Then, a unified ST-SMO method for observing tire dynamic parameters is designed. We derive the ST-SMO observer for the longitudinal tire force based on the identification of the tire effective radius. An ST-SMO algorithm for lateral tire force is also designed to individually observe lateral tire forces. Furthermore, we design the observer for tire sideslip angles. The results demonstrate that the proposed observation method for tire dynamic parameters can accurately acquire tire forces and tire sideslip angles: the observation error of longitudinal tire force is within 6.5%, the observation error of lateral tire force does not exceed 6%, the simulation observation error of tire sideslip angle is less than 3%, and the experimental observation error in the real-car test is less than 10%. The simulation error of the vertical tire force calculation method is less than 5%, and the experimental data error is around 10%.

4 Estimation of Vehicle Crucial Dynamic States

Chapter 3 investigates the estimation method, algorithm design, and experimental analysis for the observation of tire dynamic states, including the lateral tire forces and tire sideslip angles. This chapter focuses on the observation of critical states for dynamic stability, including vehicle sideslip angle and road adhesion coefficient. We design an adaptive integral correction fusion algorithm to accurately estimate vehicle sideslip angle. Within this fusion algorithm, an ASCKF method is first proposed to observe vehicle sideslip angle, with parameters adjusted adaptively based on the vehicle's dynamic states. Subsequently, a zero-reset integration method for the estimation of vehicle sideslip angle is designed. Additionally, given the complexity of vehicle driving conditions, this chapter introduces a road adhesion coefficient mode-switching observer. An RLS-based estimator and a comprehensive algorithm are proposed to perform the observation of road friction coefficient. Concretely, when the vehicle is driving or braking on a straight-line road, the RLS identification method performs the estimation task. When the vehicle performs steering maneuvers, the compensation algorithm is employed to synthesize the road adhesion coefficient. Through simulation in typical scenarios and experimental analysis of real-car tests, the effectiveness of the proposed methods for observing vehicle sideslip angle and road friction coefficient is validated.

4.1 BACKGROUND

Vehicle sideslip angle indicates vehicle corning status and shows the potential of vehicle dynamic stability [130]. Vehicle lateral tire forces may decrease greatly, and the vehicle may be out of control if the vehicle sideslip angle is abnormally large and exceeds the limitation. Therefore, vehicle sideslip angle is critical to vehicle handling and stability control. However, the commercial sensors utilized to measure vehicle sideslip angle are too expensive and cannot be widely equipped in mass-produced vehicles. Thus, the observation of vehicle sideslip angle utilizing vehicle onboard sensors needs to be studied extensively.

Many have studied the observation issue of vehicle sideslip angle, and various estimators have been proposed in recent years. In the literature [29], a Luenberger observation was designed to estimate the sideslip angle and yaw rate based on an estimation of the vehicle velocity. However, the estimation of sideslip angle is coupled together with the system uncertainties and disturbances, and those issues have also been studied [103, 147, 49]. Considering the disturbance and modeling error, H. Ohara and T. Murakami proposed

DOI: 10.1201/9781003583929-4

a novel linear sideslip angle observer and it could make the estimated error smaller than the conventional linear one by increasing the design parameter in [103]. A new approach based on combinations of the system dynamics and sensor kinematics to estimate the vehicle sideslip angle and road bank angle was proposed by [147], and it identified the uncertain vehicle parameters, such as the variations of vehicle mass and cornering stiffness of tires, using the sensor measurements. In order to work for different road surface conditions, a road-tire friction-based observation was proposed to estimate the sideslip angle by [49]. The Kalman filter was an optimal recursive estimator of the state of an uncertain dynamic system proposed by Rudolf Emil Kalman and has played a key role in many industrial applications, including the estimation of vehicle states [7]. A new variable structure EKF integrated with the sideslip angle rate feedback algorithm and damping item was proposed by [83] to compensate for the model error caused by road friction estimation inaccuracy. J. Chen et al. applied the adaptive covariance matrices technique in the UKF and combined it with the sensor-based estimator [18]. X. Jin and G. Yin designed the interacting-multiple-model-filter-based UKF and EKF to adapt multiple vehicle-road system models to variable driving conditions [73]. Some studied the multiple filter fusion algorithms and utilized other artificial intelligence algorithms, such as fuzzy modeling and neural networks, to design the sideslip angle observer. X. Li et al. proposed a hybrid decentralized filtering architecture based on the vehicle state estimator composed of two parallel EKFs [86]. A novel estimation method for the vehicle states using the EKF with a fusion algorithm was proposed by [71]. H. Du et al. first applied the Takagi-Sugeno (T-S) fuzzy modeling technique to represent the vehicle lateral dynamics with the non-linear Dugoff tire model and varying speed to design a novel sideslip angle observer [71]. Then a T-S observer was designed to simultaneously estimate the steering angle and sideslip angle with the measurements of yaw rate and vehicle speed [38]. It was designed to be robust against parameter uncertainties and unknown inputs.

Though many sideslip angle estimators so far have been designed, their performance still needs to be improved, especially when a car runs in extreme conditions. The EKF could only ensure first-order accuracy due to the linearization method utilized in it [83], and the error could be very large when the system is in its highly non-linear state. UKF is better than EKF, and it ensures second-order accuracy by the unscented transformation method, but the stability of UKF may be uncertain because it requires the prediction matrix to be positive definite [18]. The AI algorithms require comprehensive training and a lot of data support. Therefore, other new estimation algorithms still need to be designed to ensure accuracy in non-linear regions.

As we all know, the vehicle dynamics model may be inaccurate due to the friction coefficient estimator error, deviation of the vehicle and tire parameters, and so on [49, 5]. What is more, when the sideslip angle is extremely large, which means the car may be in a state of dynamic instability, tires work in nonlinear regions and the vehicle non-linearity decreases the precision of observers. However, the ESC intervenes in the brake system to ensure the car's safety based on the accuracy estimation of sideslip angle. Therefore, its estimation must be precise and still needs to be studied a lot when a car is in danger of dynamic instability.

Considering vehicle nonlinear dynamics in extreme conditions, the model error, and the measurement error caused by unknown colored noises, and due to the excellent performance of CKF, this chapter proposes a novel ASCKF estimator with the integral correction fusion and introduces a road adhesion coefficient mode-switching observer.

4.2 VEHICLE SIDESLIP ANGLE ESTIMATION

Currently, various Kalman filters are widely researched and applied in industrial control systems. The EKF algorithm can linearize nonlinear systems through Taylor expansion and is applicable to nonlinear systems, but it can only guarantee first-order accuracy. When the nonlinear system is in a strongly nonlinear state, its error can be significant. The UKF algorithm utilizes the Unscented Transformation method to ensure second-order accuracy for nonlinear systems. In terms of observation accuracy, UKF outperforms EKF. However, the stability of the UKF algorithm cannot be guaranteed because it requires the covariance matrix to be positive definite throughout the filtering process. The CKF method was first proposed by I. Arasaratnam and S. Haykin, and its core is the third-order spherical-radial cubature rule, enabling the numerical computation of the product of nonlinear functions and Gaussian density functions in nonlinear Bayesian filters. The SCKF enhances the standard CKF by incorporating the square root concept. It uses the least squares method to derive the Kalman gain and updates the covariance via matrix triangular decomposition, ensuring the covariance remains symmetric and (semi)positive definite. Therefore, considering vehicle nonlinearity in extreme operation conditions, modeling inaccuracies in the dynamic model, and measurement errors caused by unknown colored noise, this chapter adopts the SCKF algorithm to design an adaptive fusion algorithm for the estimation of vehicle sideslip angle. We propose an ASCKF method that adaptively adjusts the variance matrix based on the vehicle's dynamic states. Then, we design an estimation method using damping terms and the zero-reset integration. Consequently, the accurate vehicle sideslip angle is adaptively fused with the estimated values from the integration method.

The proposed estimation method for vehicle sideslip angle is illustrated in Fig. 4.1. The estimation method comprises four primary components: the sensor signal measurement module, the integration calculation module, the ASCKF estimation module, and the adaptive fusion module. Signals measured by onboard sensors are transmitted to other calculation modules. The covariance matrix in the ASCKF algorithm can be adaptively adjusted based on vehicle dynamic states. Simultaneously, an integration method incorporating damping terms and zero-reset is utilized to estimate the vehicle sideslip angle. Finally, the adaptive fusion module determines the weights of the estimated values from the integration method and the ASCKF algorithm to achieve an accurate observation.

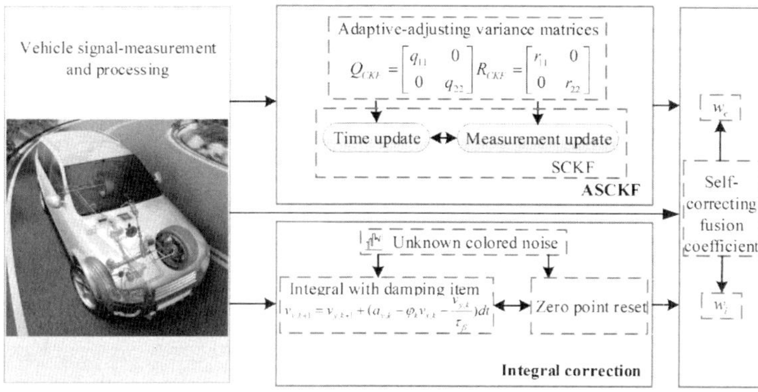

Figure 4.1 Algorithm architecture for observing vehicle sideslip angle.

4.2.1 BASIC PRINCIPLES OF CUBATURE KALMAN FILTER

The core of the CKF algorithm is to derive the filtering formula for nonlinear Gaussian systems based on Bayesian estimation and the minimum variance criterion. It then approximates the posterior mean and covariance of the system states using suitable approximation methods.

4.2.1.1 Bayesian Filtering Estimation

For a general nonlinear Gaussian system, as shown in Eq. (4.1), where the system state and measurement vectors are denoted as $\boldsymbol{x}_{bs,k}$ and $\boldsymbol{z}_{bs,k}$, respectively. $f_{bs}(\cdot)$ and $h_{bs}(\cdot)$ represent the system state transition and measurement functions. $\boldsymbol{v}_{bs,k}$ and $\boldsymbol{w}_{bs,k}$ are mutually independent Gaussian white noises with zero mean, representing the system noise and measurement noise, with variances denoted by $Q_{bs,k}$ and $R_{bs,k}$, respectively.

$$\begin{cases} \boldsymbol{x}_{bs,k+1} = f_{bs}(\boldsymbol{x}_{bs,k}, \boldsymbol{u}_{bs,k}) + \boldsymbol{v}_{bs,k} \\ \boldsymbol{z}_{bs,k} = h_{bs}(\boldsymbol{x}_{bs,k}, \boldsymbol{u}_{bs,k}) + \boldsymbol{w}_{bs,k} \end{cases} \tag{4.1}$$

The core concept of Bayesian filtering estimation is to solve the posterior probability density function, which comprehensively describes the state estimation of the nonlinear system, using the information obtained from system measurements. We can effectively estimate the system states by utilizing the state probability density function, which describes the statistical characteristics of the system, and integrating suitable approximation methods and estimation criteria. Thus, recursive Bayesian filtering estimation calculates $p_{bs}(\boldsymbol{x}_{k+1}|\boldsymbol{D}_{bs,k+1})$ based on the known posterior probability density function $p_{bs}(\boldsymbol{x}_{bs,k}|\boldsymbol{D}_{bs,k})$ at time k, in conjunction with the measurement $\boldsymbol{z}_{bs,k+1}$ at time $k+1$. Two main procedures are involved: the time update and measurement update, as shown in Fig. 4.2, where $\boldsymbol{D}_{bs,k} = \{\boldsymbol{z}_{bs,1}, \boldsymbol{u}_{bs,1}, \boldsymbol{z}_{bs,2}, \boldsymbol{u}_{bs,2}, \ldots, \boldsymbol{z}_{bs,k}, \boldsymbol{u}_{bs,k}\}$. The state probability density function for the one-step time update can be computed using Eq. (4.2), and the state posterior probability density function at time $k+1$ is calculated by using Eqs. (4.3) and (4.4) [68].

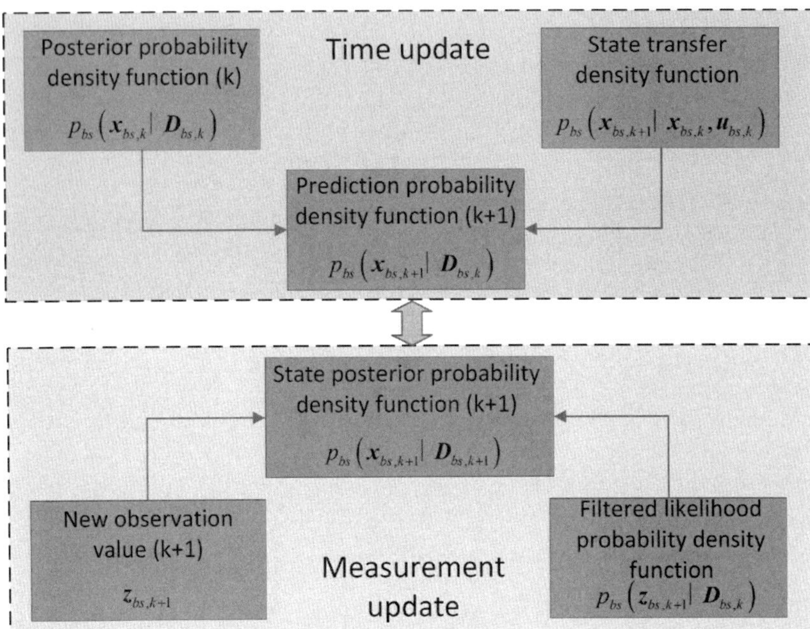

Figure 4.2 The flowchart of recursive Bayesian filtering estimation.

$$p_{bs}(\boldsymbol{x}_{bs,k+1}|\boldsymbol{D}_{bs,k}) = \int p_{bs}(\boldsymbol{x}_{bs,k}|\boldsymbol{D}_{bs,k}) p_{bs}(\boldsymbol{x}_{bs,k+1}|\boldsymbol{x}_{bs,k}, \boldsymbol{u}_{bs,k}) \mathrm{d}\boldsymbol{x}_{bs,k} \tag{4.2}$$

$$p_{bs}(\boldsymbol{x}_{bs,k+1}|\boldsymbol{D}_{bs,k+1}) = \frac{1}{c_{bs,k+1}} p_{bs}(\boldsymbol{x}_{bs,k+1}|\boldsymbol{D}_{bs,k}) p_{bs}(z_{bs,k+1}|\boldsymbol{x}_{bs,k+1}, \boldsymbol{u}_{bs,k+1}) \tag{4.3}$$

The normalization constant $c_{bs,k+1}$ is computed as follows:

$$\begin{aligned} c_{bs,k+1} &= p_{bs}(\boldsymbol{z}_{bs,k+1}|\boldsymbol{D}_{bs,k}, \boldsymbol{x}_{bs,k+1}) \\ &= \int p_{bs}(\boldsymbol{x}_{bs,k+1}|\boldsymbol{D}_{bs,k}) p_{bs}(\boldsymbol{z}_{bs,k+1}|\boldsymbol{x}_{bs,k+1}, \boldsymbol{u}_{bs,k+1}) \mathrm{d}\boldsymbol{x}_{bs,k+1} \end{aligned} \tag{4.4}$$

Given that the actual random phenomena can be described by using Gaussian processes, both the prior probability density function $p_{bs}(\boldsymbol{x}_{bs,k+1}|\boldsymbol{D}_{bs,k})$ and the filtering likelihood probability density function $p_{bs}(\boldsymbol{z}_{bs,k+1}|\boldsymbol{D}_{bs,k})$ follow Gaussian distributions. Thus, the posterior probability density function $p_{bs}(\boldsymbol{x}_{bs,k+1}|\boldsymbol{D}_{bs,k+1})$ also follows a Gaussian distribution, as shown in Eqs. (4.5) and (4.6).

$$p_{bs}(\boldsymbol{x}_{bs,k}|\boldsymbol{D}_{bs,k}) = N(\boldsymbol{x}_{bs,k};\hat{\boldsymbol{x}}_{bs,k|k},\boldsymbol{P}_{bs,k|k}) \tag{4.5}$$

$$p_{bs}(\boldsymbol{z}_{bs,k+1}|\boldsymbol{D}_{bs,k}) = N(\boldsymbol{z}_{bs,k+1};\hat{\boldsymbol{z}}_{bs,k+1|k},\boldsymbol{P}_{bs,zz,k+1|k}) \tag{4.6}$$

Thus, through the procedure of time update, we can calculate the mean value and covariance of the system state $\hat{\boldsymbol{x}}_{bs,k+1|k}$ and $\boldsymbol{P}_{bs,k+1|k}$ at time $k+1$ given time k, which is expressed as

$$\begin{aligned}\hat{\boldsymbol{x}}_{bs,k+1|k} &= \mathbb{E}\left[f_{bs}(\boldsymbol{x}_{bs,k},\boldsymbol{u}_{bs,k})|\boldsymbol{D}_{bs,k}\right]\\ &= \int f_{bs}(\boldsymbol{x}_{bs,k},\boldsymbol{u}_{bs,k})p_{bs}(\boldsymbol{x}_{bs,k}|\boldsymbol{D}_{bs,k})\mathrm{d}\boldsymbol{x}_{bs,k}\\ &= \int f_{bs}(\boldsymbol{x}_{bs,k},\boldsymbol{u}_{bs,k})N(\boldsymbol{x}_{bs,k};\hat{\boldsymbol{x}}_{bs,k|k},\boldsymbol{P}_{bs,k|k})\mathrm{d}\boldsymbol{x}_{bs,k}\end{aligned} \tag{4.7}$$

$$\begin{aligned}\boldsymbol{P}_{bs,k+1|k} &= \mathbb{E}\left[(\boldsymbol{x}_{bs,k+1}-\hat{\boldsymbol{x}}_{bs,k+1|k})(\boldsymbol{x}_{bs,k+1}-\hat{\boldsymbol{x}}_{bs,k+1|k})^T|\boldsymbol{D}_{bs,k}\right]\\ &= \int f_{bs}(\boldsymbol{x}_{bs,k},\boldsymbol{u}_{bs,k})f_{bs}^T(\boldsymbol{x}_{bs,k},\boldsymbol{u}_{bs,k})N(\boldsymbol{x}_{bs,k};\hat{\boldsymbol{x}}_{bs,k|k},\boldsymbol{P}_{bs,k|k})\mathrm{d}\boldsymbol{x}_{bs,k}\\ &\quad -\hat{\boldsymbol{x}}_{bs,k+1|k}\hat{\boldsymbol{x}}_{bs,k+1|k}^T+\boldsymbol{Q}_{bs,k}\end{aligned} \tag{4.8}$$

In the procedure of measurement update, the predicted value $\hat{\boldsymbol{z}}_{bs,k+1|k}$ and its covariance matrix $\boldsymbol{P}_{bs,zz,k+1|k}$ for the measurement at time $k+1$ given time k in Eq. (4.6) are calculated as follows:

$$\hat{\boldsymbol{z}}_{bs,k+1|k} = \int h_{bs}(\boldsymbol{x}_{bs,k+1},\boldsymbol{u}_{bs,k+1})N(\boldsymbol{x}_{bs,k+1};\hat{\boldsymbol{x}}_{bs,k+1|k},\boldsymbol{P}_{bs,k+1|k})\mathrm{d}\boldsymbol{x}_{bs,k+1} \tag{4.9}$$

$$\begin{aligned}\boldsymbol{P}_{bs,zz,k+1|k} &= \int h_{bs}(\boldsymbol{x}_{bs,k+1},\boldsymbol{u}_{bs,k+1})h_{bs}^T(\boldsymbol{x}_{bs,k+1},\boldsymbol{u}_{bs,k+1})\\ &\quad\cdot N(\boldsymbol{x}_{bs,k+1};\hat{\boldsymbol{x}}_{bs,k+1|k},\boldsymbol{P}_{bs,k+1|k})\mathrm{d}\boldsymbol{x}_{bs,k+1}-\hat{\boldsymbol{z}}_{bs,k+1|k}\hat{\boldsymbol{z}}_{bs,k+1|k}^T\\ &\quad+\boldsymbol{R}_{bs,k+1}\end{aligned} \tag{4.10}$$

In addition, we should also compute the cross-covariance matrix, which can be calculated as shown in Eq. (4.11), to yield the state posterior probability density function.

$$\begin{aligned}\boldsymbol{P}_{bs,xz,k+1|k} &= \mathbb{E}\left[(\boldsymbol{x}_{bs,k+1}-\hat{\boldsymbol{x}}_{bs,k+1|k})(\boldsymbol{z}_{bs,k+1}-\hat{\boldsymbol{z}}_{bs,k+1|k})^T|\boldsymbol{D}_{bs,k}\right]\\ &= \int \boldsymbol{x}_{bs,k+1}h_{bs}^T(\boldsymbol{x}_{bs,k+1},\boldsymbol{u}_{bs,k+1})N(\boldsymbol{x}_{bs,k+1};\hat{\boldsymbol{x}}_{bs,k+1|k},\boldsymbol{P}_{bs,k+1|k})\\ &\quad\times \mathrm{d}\boldsymbol{x}_{bs,k+1}-\hat{\boldsymbol{x}}_{bs,k+1|k}\hat{\boldsymbol{z}}_{bs,k+1|k}^T\end{aligned} \tag{4.11}$$

Therefore, at each sampling time-step, we first obtain new measurement $z_{bs,k+1}$. Then, we apply the minimum variance criterion and the principles of Bayesian filtering estimation to yield the optimal Gaussian filter for the nonlinear system (Shown in Eq. (4.1)) as follows.

$$p_{bs}(\boldsymbol{x}_{bs,k+1}|\boldsymbol{D}_{bs,k+1}) = N(\boldsymbol{x}_{bs,k+1}; \hat{\boldsymbol{x}}_{bs,k+1|k+1}, \boldsymbol{P}_{bs,k+1|k+1}) \tag{4.12}$$

$$\hat{\boldsymbol{x}}_{bs,k+1|k+1} = \hat{\boldsymbol{x}}_{bs,k+1|k} + \boldsymbol{K}_{bs,k+1}(\boldsymbol{z}_{bs,k+1} - \hat{\boldsymbol{z}}_{bs,k+1|k}) \tag{4.13}$$

$$\boldsymbol{P}_{bs,k+1|k+1} = \boldsymbol{P}_{bs,k+1|k} - \boldsymbol{K}_{bs,k+1}\boldsymbol{P}_{bs,zz,k+1|k}\boldsymbol{K}^T_{bs,k+1} \tag{4.14}$$

$$\boldsymbol{K}_{bs,k+1} = \boldsymbol{P}_{bs,xz,k+1|k}\boldsymbol{P}^{-1}_{bs,zz,k+1|k} \tag{4.15}$$

As seen in Eqs. (4.12) to (4.15), the integrals presented in Eqs. (4.7) to (4.11) are the fundamental components, involving the product of a nonlinear function and a Gaussian probability density function as presented in Eq. (4.16). Such Gaussian integrals are challenging to solve analytically and require numerical integration methods to compute numerical approximations.

$$I_g(f_g) = \int f_g(\boldsymbol{x}_g)\exp\left(-\boldsymbol{x}_g\boldsymbol{x}_g^T\right)\mathrm{d}\boldsymbol{x}_g \tag{4.16}$$

4.2.1.2 Third-Order Spherical-Radial Cubature Rule

The core problem of Bayesian filtering estimation, as presented in Eq. (4.16), can be reformulated into a spherical-radial integral. This integral can then be approximated using cubature rules [5].

Given that $\boldsymbol{x}_g = r_g\boldsymbol{y}_g$, and $\boldsymbol{y}_g^T\boldsymbol{y}_g = 1$, we can yield $\boldsymbol{x}_g^T\boldsymbol{x}_g = r_g^2$ with $r \in [0, \infty)$. Thus, the integral of the product of the Gaussian probability density function and the nonlinear function, as shown in Eq. (4.16), can be transformed into a double integral in spherical-radial coordinates:

$$I_g(f_g) = \int_0^{\infty}\int_{U_n} f_g(r_g\boldsymbol{y}g)rg^{n-1}\exp\left(-r^2\right)\mathrm{d}\sigma(\boldsymbol{y}_g)\mathrm{d}r_g \tag{4.17}$$

where $U_n = \boldsymbol{y}g \in \mathbb{R}^n|\boldsymbol{y}^Tg\boldsymbol{y}_g = 1$ represents the unit sphere, and $\sigma(\cdot)$ is the surface measure. Further, the above integral can be decomposed into a radial integral and a spherical integral:

$$R_g = \int_0^{\infty} S_g(r_g)r_g^{n-1}\exp\left(-r_g^2\right)\mathrm{d}r_g \tag{4.18}$$

$$S_g(r_g) = \int_{U_n} f_g(r_g\boldsymbol{y}_g)\mathrm{d}\sigma(\boldsymbol{y}_g) \tag{4.19}$$

Let $x_r = r_g^2$, the radial integral shown in Eq. (4.18) can be approximated using the Gauss-Laguerre integration rule:

$$\begin{aligned} R_g &= \frac{1}{2}\int_0^{\infty} S_g(\sqrt{x_r})x_r^{n/2-1}\exp(-x_r)\mathrm{d}x_r \\ &\approx \frac{1}{2}\cdot S_g\left(\sqrt{n/2}\right)\cdot\int_0^{\infty} x_r^{n/2-1}\exp(-x_r)\mathrm{d}x_r \end{aligned} \tag{4.20}$$

The spherical integral shown in Eq. (4.19) can be approximated using the third-order spherical cubature rule:

$$S_g(r_g) \approx \sum_{i=1}^{2n} \omega_{g,i} f_g(r_g s_{g,i}) \tag{4.21}$$

where $s_{g,i} = [\![1]\!]_i$ represents the set of all symmetrically distributed points on the intersection of the n-dimensional unit sphere and the coordinate axes. Each point is assigned an equal weight, as specified in Eqs. (4.22) and (4.23).

$$[\![1]\!] = \left\{ \begin{bmatrix} 1 \\ 0 \end{bmatrix} \begin{bmatrix} 0 \\ 1 \end{bmatrix} \begin{bmatrix} -1 \\ 0 \end{bmatrix} \begin{bmatrix} 0 \\ -1 \end{bmatrix} \right\} \tag{4.22}$$

$$\omega_{g,i} = \frac{1}{2n} \frac{2\sqrt{\pi^n}}{\int_0^\infty x_r^{n/2-1} \exp(-x_r) \mathrm{d}x_r} \tag{4.23}$$

Combining Eqs. (4.20) and (4.21), we can obtain the numerical approximation of the Gaussian integral using the third-order spherical-radial cubature rule, which is expressed as:

$$I_g(f_g) \approx \frac{\sqrt{\pi^n}}{2n} \sum_{i=1}^{2n} f_g\left(\sqrt{\frac{n}{2}} [\![1]\!]_i\right) \tag{4.24}$$

4.2.2 VEHICLE SIDESLIP ANGLE SCKF ESTIMATOR

In the proposed SCKF algorithm, we use the four-wheel 7DOF vehicle dynamic model introduced in Section 2.4, which is discretized using Euler's method:

$$\begin{cases} \boldsymbol{X}_{\beta,k+1} = f_\beta(\boldsymbol{X}_{\beta,k}, \boldsymbol{u}_{\beta,k}) + \boldsymbol{v}_{\beta,k} \\ \boldsymbol{Z}_{\beta,k} = h_\beta(\boldsymbol{X}_{\beta,k}, \boldsymbol{u}_{\beta,k}) + \boldsymbol{w}_{\beta,k} \end{cases} \tag{4.25}$$

where $\boldsymbol{X}_{\beta,k+1} = [v_{y,k+1}, \varphi_{k+1}]^T$ and $\boldsymbol{Z}_{\beta,k} = [a_{y,k}, \varphi_k]^T$ represent the system state and measurement state in the SCKF observer, respectively. $\boldsymbol{v}_{\beta,k}$ and $\boldsymbol{w}_{\beta,k}$ denote the model noise and measurement noise of the observation system, respectively. The detailed calculation method for Eq. (4.25) is as follows:

$$\begin{bmatrix} v_{y,k+1} \\ \varphi_{k+1} \end{bmatrix} = \begin{bmatrix} v_{y,k} + [F_{yrl,k} + F_{yrr,k} - m v_{x,k} \varphi_k + (F_{yrl,k} \\ + F_{yrr,k}) \cos\delta_k + (F_{xfl,k} + F_{xfr,k}) \sin\delta_k] \mathrm{d}t/m; \\ \varphi_k + [(F_{yfl,k} + F_{yfr,k}) l_f \cos\delta_k + 0.5(F_{yfl,k} - F_{yfr,k}) T_{wb} \sin\delta_k \\ -(F_{yrl,k} + F_{yrr,k}) l_r + (F_{xfl,k} + F_{xfr,k}) l_f \sin\delta_k - 0.5(F_{xfl,k} \\ -F_{xfr,k}) T_{wb} \cos\delta_k - 0.5(F_{xrl,k} - F_{xrr,k}) T_{wb}] \mathrm{d}t/I_z \end{bmatrix} \tag{4.26}$$

The iterative calculation steps of the SCKF algorithm are as follows [5]:

Firstly, we calculate the cubature points $\xi_{SCKF,i}$ and their corresponding weights $\omega_{SCKF,i}$as follows:

$$\begin{cases} \xi_{\mathrm{SCKF,i}} = \sqrt{\frac{c}{2}} [\![1]\!]_{\mathrm{i}}, \\ \omega_{\mathrm{SCKF,i}} = \frac{1}{c}, \quad i = 1, 2, ..., c. \end{cases} \tag{4.27}$$

where $c = 2n_{SCKF}$, with n_{SCKF} denoting the state dimension of the SCKF algorithm. The set $[\![1]\!]$ comprises the complete combinations of the unit vector $e_{cp} = [1,0,0,...,0]^T$, and $[\![1]\!]_i$ represents the i-th point in the set $[\![1]\!]$.

(1) Time Update

1) Compute the cubature points of the SCKF observer ($i = 1,2,...,c.$):

$$\boldsymbol{X}_{\beta,i,k-1|k-1} = \boldsymbol{S}_{\beta,k-1|k-1}\xi_{SCKF,i} + \hat{\boldsymbol{x}}_{\beta,k-1|k-1} \tag{4.28}$$

2) Calculate the cubature points propagated through the observer system equations ($i = 1,2,...,c.$):

$$\boldsymbol{X}^*_{\beta,i,k|k-1} = f_\beta(\boldsymbol{X}_{\beta,i,k-1|k-1}, \boldsymbol{u}_{\beta,k-1}) \tag{4.29}$$

3) Predict the state at the next time-step:

$$\hat{\boldsymbol{x}}_{\beta,k|k-1} = \frac{1}{c}\sum_{i=1}^{c}\boldsymbol{X}^*_{\beta,i,k|k-1} \tag{4.30}$$

4) Predict the square root covariance factor of the state at the next time-step:

$$\boldsymbol{S}_{\beta,k|k-1} = \text{Tria}([\boldsymbol{\chi}^*_{\beta,k|k-1} \quad \boldsymbol{S}_{\beta,Q,k-1}]) \tag{4.31}$$

where $\boldsymbol{S}_{\beta,Q,k-1}$ represents the square root factor of $\boldsymbol{Q}_{CFK,k-1}$ such that $\boldsymbol{Q}_{CKF,k-1} = \boldsymbol{S}_{\beta,Q,k-1}\boldsymbol{S}^T_{\beta,Q,k-1}$. $\boldsymbol{S}_{tria} = \text{Tria}(\boldsymbol{\Theta})$ represents a common triangularization algorithm, where $\boldsymbol{S}_{tria}$ is a lower triangular matrix. Given that $\Re$ is the upper triangular matrix obtained by the factorization of $\boldsymbol{S}_{tria}$, we can yield that $\boldsymbol{S}_{tria} = \Re^T$.

5) The method for calculating the weighted central matrix (subtracting the prior mean) is as follows:

$$\boldsymbol{\chi}^*_{\beta,k|k-1} = \frac{1}{\sqrt{c}}\left[\boldsymbol{X}^*_{\beta,1,k|k-1} - \hat{x}_{\beta,k|k-1} \quad \cdots \quad \cdots \quad \boldsymbol{X}^*_{\beta,c,k|k-1} - \hat{x}_{\beta,k|k-1}\right] \tag{4.32}$$

(2) Measurement Update

1) Compute the values of the cubature points ($i = 1,2,...,c.$):

$$\boldsymbol{X}_{\beta,i,k|k-1} = \boldsymbol{S}_{\beta,k|k\_1}\xi_{SCKF,i} + \hat{\boldsymbol{x}}_{\beta,k|k-1} \tag{4.33}$$

2) Calculate the cubature points propagated through the observer measurement equations ($i = 1,2,...,c.$):

$$\boldsymbol{Z}_{\beta,i,k|k-1} = h_\beta(\boldsymbol{X}_{\beta,i,k|k-1}, \boldsymbol{u}_{\beta,k}) \tag{4.34}$$

3) Predict the measurement value at the next time-step:

$$\hat{\boldsymbol{z}}_{\beta,k|k-1} = \frac{1}{c}\sum_{i=1}^{c} \boldsymbol{Z}_{\beta,i,k|k-1} \tag{4.35}$$

4) Estimate the square root factor of the updated covariance matrix:

$$\boldsymbol{S}_{zz,k|k-1} = \text{Tria}([\ \boldsymbol{Z}_{\beta,k|k-1} \quad \boldsymbol{S}_{\beta,R,k}\ ]) \tag{4.36}$$

where $\boldsymbol{S}_{\beta,R,k}$ represents the square root factor of $\boldsymbol{R}_{CKF,k}$ such that $\boldsymbol{R}_{CKF,k} = \boldsymbol{S}_{\beta,R,k}\boldsymbol{S}_{\beta,R,k}^T$.

5) Calculate the weighted central matrix:

$$\boldsymbol{Z}_{\beta,k|k-1} = \frac{1}{\sqrt{c}}[\ \boldsymbol{Z}_{\beta,1,k|k-1} - \hat{\boldsymbol{z}}_{\beta,k|k-1} \quad \cdots \quad \cdots \quad \boldsymbol{Z}_{\beta,c,k|k-1} - \hat{\boldsymbol{z}}_{\beta,k|k-1}\ ] \tag{4.37}$$

6) Estimate the cross-covariance matrix of the prediction:

$$\boldsymbol{P}_{xz,k|k-1} = \boldsymbol{\chi}_{\beta,k|k-1}\boldsymbol{Z}_{\beta,k|k-1}^T \tag{4.38}$$

where the weighted central matrix is given by:

$$\boldsymbol{\chi}_{\beta,k|k-1} = \frac{1}{\sqrt{c}}[\ \boldsymbol{X}_{\beta,1,k|k-1} - \hat{\boldsymbol{x}}_{\beta,k|k-1} \quad \cdots \quad \cdots \quad \boldsymbol{X}_{\beta,c,k|k-1} - \hat{\boldsymbol{x}}_{\beta,k|k-1}\ ] \tag{4.39}$$

7) Estimate the filtering gain:

$$\boldsymbol{W}_{CKF,k} = \frac{(\boldsymbol{P}_{xz,k|k-1}/\boldsymbol{S}_{zz,k|k-1}^T)}{\boldsymbol{S}_{zz,k|k-1}} \tag{4.40}$$

where the symbol / denotes the matrix right division operation. For instance, $\boldsymbol{M}_{ins}/\boldsymbol{N}_{ins}$ employs back-substitution for the upper triangular matrix $\boldsymbol{N}_{ins}$ and forward-substitution for the lower triangular matrix $\boldsymbol{N}_{ins}$.

8) Compute the updated state:

$$\hat{\boldsymbol{x}}_{\beta,k|k} = \hat{\boldsymbol{x}}_{\beta,k|k-1} + \boldsymbol{W}_{CKF,k}(\boldsymbol{z}_{\beta,k} - \hat{\boldsymbol{z}}_{\beta,k|k-1}) \tag{4.41}$$

9) Calculate the corresponding covariance square root factor of the state error:

$$\boldsymbol{S}_{\beta,k|k} = \text{Tria}([\boldsymbol{\chi}_{\beta,k|k-1} - \boldsymbol{W}_{CKF,k}\boldsymbol{Z}_{\beta,k|k-1} \quad \boldsymbol{W}_{CKF,k}\boldsymbol{S}_{\beta,R,k}]) \tag{4.42}$$

4.2.3 ADAPTIVE ADJUSTMENT OF ASCKF VARIANCE MATRIX

In numerous research studies, the variance matrix designed for Kalman filtering observers is typically set to a constant value. However, during the observation process, the system's model noise and measurement noise are often unknown. To address this, this study proposes the ASCKF-based observer that adaptively adjusts the variance matrix based on vehicle dynamic states, including lateral acceleration a_y, yaw rate φ, and tire slip ratio λ_{ij}. This algorithm achieves higher accuracy compared to traditional Kalman filter-based observers.

When vehicle lateral tire forces are large, the vehicle dynamics exhibits strong nonlinearity, leading to decreased accuracy of the vehicle dynamic model. If a_y is large, the accuracy of the vehicle dynamic model may decrease, and consequently, the confidence in the observation values of the vehicle dynamic model decreases. Therefore, the variance of the observer system model noise is adaptively adjusted to be positively correlated with a_y, φ, and λ_{ij}. The larger the measured value of the dynamic state, the higher its confidence. Conversely, when a_y is small, the confidence in the measurement value decreases, and the reliability of the integral result obtained from a_y also decreases. Thus, the measurement noise variance is inversely correlated with a_y and φ and is adapted accordingly. In the Kalman filter, the confidence in the system model and measurement results is determined by the covariance matrix of the filter noise. By adjusting the covariance matrix value, the filter can be adaptively adjusted. The design of the noise covariance matrix is as follows:

$$\boldsymbol{Q}_{CKF} = \begin{bmatrix} \begin{array}{l} c_{q11}\left|a_y\right| + c_{q12}(\sum\limits_{i,j}\lambda_{ij}) \\ +c_{q13}\left|\varphi\right| + c_{q14} \end{array} & 0 \\ 0 & c_{q21}\left|a_y\right| + c_{q22}(\sum\limits_{i,j}\lambda_{ij}) + c_{q23}\left|\varphi\right| + c_{q24} \end{bmatrix} \tag{4.43}$$

$$\boldsymbol{R}_{CKF} = \begin{bmatrix} (1+c_{r11})[c_{r12}(c_{r13}-\left|a_y\right|) + c_{r14}(c_{r15}-\left|\varphi\right|)] & 0 \\ 0 & c_{r21} \end{bmatrix} \tag{4.44}$$

where $c_{q11}, c_{q12}, c_{q13}, c_{q14}, c_{q21}, c_{q22}, c_{q23}, c_{q24}, c_{r11}, c_{r12}, c_{r13}, c_{r14}, c_{r15}$, and c_{r21} can be calibrated by the results of simulations and real-car experiments.

4.2.4 INTEGRATION CORRECTION FUSION

Kalman filtering-based observers cannot avoid model errors caused by inaccurate vehicle dynamic models. Conversely, the integration method is derived from real-time signals of the onboard gyroscope without using a vehicle dynamic model. Therefore, the integration method can correct model errors in Kalman-filtering-based observers. Additionally, due to the presence of unknown colored noise during the measurement process of onboard sensors, errors are inevitable in the integration method based on sensor signals. To mitigate these errors caused by the sensor signal noise, we incorporate damping terms and design a zero-reset method.

4.2.4.1 Integration Calculation with Damping Term

According to the vehicle dynamic model described in Section 2.4, signals such as lateral acceleration measured by vehicle-mounted sensors can be integrated to calculate the vehicle's lateral velocity and vehicle sideslip angle as follows:

$$v_y = \int (a_y - \varphi v_x)\mathrm{d}t \tag{4.45}$$

$$\beta = \arctan\left(\frac{v_y}{v_x}\right) \approx \frac{v_y}{v_x} \tag{4.46}$$

To reduce offset in sensor signals and cumulative errors caused by colored noise, a damping coefficient τ_β is designed to correct the integral term, which is expressed as follows:

$$v_y = \int \left(a_y - \varphi v_x - \frac{v_y}{\tau_\beta}\right)\mathrm{d}t \tag{4.47}$$

To reduce cumulative errors caused by the sensor signal noise, the damping coefficient τ_β in the above equation must be bounded, and in this algorithm, its value is set to be greater than 1.

The discrete expression of Eq. 4.47 and the integrated value of the vehicle sideslip angle can be computed as follows:

$$\begin{cases} v_{y,k+1} = v_{y,k} + \left(a_{y,k} - \varphi_k v_{x,k} - \dfrac{v_{y,k}}{\tau_\beta}\right) dt \\ \beta_k = \arctan\left(\dfrac{v_{y,k}}{v_{x,k}}\right) \approx \dfrac{v_{y,k}}{v_{x,k}} \end{cases} \tag{4.48}$$

4.2.4.2 Zero-Reset Method

The core of the zero-reset method lies in the design of zero-point resetting. This chapter employs the first-order delay of the yaw rate signal to reset the zero point, which is based on the response behavior of yaw rate and vehicle sideslip angle depicted by the single-track dynamic model as introduced in Fig. 2.9.

First, we transform the single-track 2DOF dynamic model as described in Section 2.4 into the following form:

$$\begin{bmatrix} \dot{\beta}_{bm} \\ \dot{\varphi}_{bm} \end{bmatrix} = \begin{bmatrix} A_{bm1} & A_{bm2} \\ A_{bm3} & A_{bm4} \end{bmatrix} \begin{bmatrix} \beta_{bm} \\ \varphi_{bm} \end{bmatrix} + \begin{bmatrix} B_{bm1} \\ B_{bm2} \end{bmatrix} \delta \tag{4.49}$$

where

$$\begin{cases} A_{bm1} = \dfrac{k_f + k_r}{mv_x}, A_{bm2} = \dfrac{l_f k_f - l_r k_r}{mv_x^2} - 1, B_{bm1} = -\dfrac{k_f}{mv_x} \\ A_{bm3} = \dfrac{l_f k_f - l_r k_r}{I_z}, A_{bm4} = \dfrac{l_f^2 k_f + l_r^2 k_r}{v_x I_z}, B_{bm2} = -\dfrac{l_f k_f}{I_z} \end{cases} \tag{4.50}$$

Define the Laplace transform factor $j\Omega = s_{bm}$, then we can transform the Eq. (4.49) into the Laplace form, which is expressed by

$$\begin{aligned} (s_{bm}^2 - T_{bmA}s_{bm} + D_{bmA})\beta_{bm}(s_{bm}) &= (B_{bm1}s_{bm} - B_{bm1}A_{bm4} + B_{bm2}A_{bm2})\delta \\ (s_{bm}^2 - T_{bmA}s_{bm} + D_{bmA})\varphi_{bm}(s_{bm}) &= (B_{bm2}s_{bm} - B_{bm2}A_{bm1} + B_{bm1}A_{bm3})\delta \end{aligned} \tag{4.51}$$

where $T_{bmA} = A_{bm1} + A_{bm4}$, $D_{bmA} = A_{bm1}A_{bm4} - A_{bm2}A_{bm3}$.

Based on the Laplace form of the 2DOF dynamic model, as shown in Eq. (4.51), we can yield the phase angles of vehicle sideslip angle β and yaw rate φ, which are computed as follows:

$$\begin{cases} \psi_{\beta_{bm}} = \arctan\dfrac{B_{bm1}\Omega}{B_{bm2}A_{bm2} - B_{bm1}A_{bm4}} - \arctan\dfrac{-T_{bmA}\Omega}{D_{bmA} - \Omega^2} \\ \psi_{\varphi_{bm}} = \arctan\dfrac{B_{bm2}\Omega}{B_{bm1}A_{bm3} - B_{bm2}A_{bm1}} - \arctan\dfrac{-T_{bmA}\Omega}{D_{bmA} - \Omega^2} \end{cases} \tag{4.52}$$

Thus, based on $\psi_{\beta_{bm}}$ and $\psi_{\varphi_{bm}}$, we can yield:

$$\mathrm{d}T = \frac{\psi_{\varphi_{bm}} - \psi_{\beta_{bm}}}{\Omega} \tag{4.53}$$

Finally, we process the measured signal of vehicle yaw rate using $\mathrm{d}T$ to obtain the zero-reset point for the integration method. When vehicle dynamic nonlinearity is sufficiently small and the value calculated by the integration method is minimal, the zero-reset point is determined by the first-order delay of the measured yaw rate signal, which corresponds to the zero of $1/(\mathrm{d}T \cdot s + 1) \cdot \varphi$.

4.2.4.3 Adaptive Fusion

We define w_{est} as the adaptive weight of the estimated value from the AS-CKF algorithm and w_{int} as the weight of the calculated value based on the integration of onboard sensor signals. The adaptive fusion factors w_{est} and w_{int} decrease as the errors in their respective observation values increase, ensuring the accuracy of the fused observations. When vehicle sideslip angle is small and the vehicle is in its linear dynamic region, the error of the dynamic model is minimal, but the noise from the onboard gyroscope signal constitutes a larger portion of the measurement signal. In this case, w_{int} should be small and w_{est} should be large. Conversely, when the vehicle is in a highly nonlinear dynamic state, w_{int} should increase while w_{est} should decrease.

The first-order delay of the derivative of vehicle lateral velocity, Γ_{wT}, represents the degree of vehicle dynamic nonlinearity. Consequently, the formula for calculating the adaptive fusion factors is as follows:

$$\begin{cases} w_{int} = \min(\chi_{wlimit}, \kappa_w(\Gamma_{wT}/\mu)^2) \\ w_{est} = 1 - \min(\chi_{wlimit}, \kappa_w(\Gamma_{wT}/\mu)^2) \end{cases} \tag{4.54}$$

where χ_{wlimit} is the upper limit of w_{int}, which is set to 0.95 in this book. κ_w is the proportion factor for calculating the adaptive fusion factors, which is calibrated through the simulation data and real-car experimental data. The acquisition of the road friction coefficient will be discussed in the next section.

Therefore, through the adaptive fusion of the estimated value from the ASCKF algorithm and the integration method, an accurate observation of vehicle sideslip angle is obtained as follows:

$$\hat{\beta} = \hat{\beta}_{est} \cdot w_{est} + \hat{\beta}_{int} \cdot w_{int} \tag{4.55}$$

where $\hat{\beta}_{est}$ is the value observed by the proposed ASCKF, and $\hat{\beta}_{int}$ is the value calculated based on the zero-reset integration method.

4.3 ROAD FRICTION COEFFICIENT OBSERVATION

The accurate estimation of road friction coefficient has always been a key issue of vehicle dynamic stability control. As a critical state parameter, the road adhesion coefficient determines the peak values of tire forces, thereby determining vehicle dynamic stability boundary. Various maneuvers are required to be performed to handle complex and variable traffic scenarios, which makes it challenging to estimate road friction coefficient accurately. To address this, considering diverse driving scenarios such as straight-line driving, straight-line braking, and steering, this section explores an estimation algorithm for road friction coefficient and proposes a mode-switching approach. When a vehicle performs driving or braking maneuvers, the mode-switching observer automatically switches to an estimation method based on longitudinal dynamics. For steering scenarios, a comprehensive estimation algorithm based on the nonlinearity of vehicle dynamics is employed. This adaptive mode-switching enables accurate road friction coefficient observation under all driving conditions.

4.3.1 ESTIMATION OF ROAD FRICTION COEFFICIENT BASED ON LONGITUDINAL DYNAMICS

When the vehicle performs straight-line driving or straight-line braking maneuvers, the mode-switching observer automatically switches to estimating the road friction coefficient based on longitudinal dynamics. This process involves calculating the friction coefficient using the tire slip ratio and tire force, derived from wheel speed and vehicle speed. By analyzing the μ-slip curve, we can identify the road friction coefficient. In varying road adhesion conditions, the longitudinal force applied to the vehicle is proportional to both the slip ratio and the vertical force. This relationship can be described as follows [54].

$$F_x = \kappa_\mu \cdot \lambda \cdot F_z \tag{4.56}$$

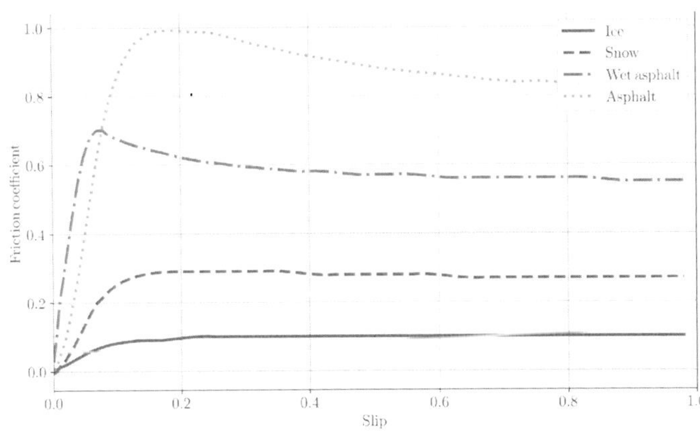

Figure 4.3 Demonstration of μ-slip curves on different road surfaces [54]

As shown in Fig. 4.3, the slope of the linear part of the μ-slip curve increases with the road adhesion coefficient. Thus, observing the slope of the linear part of the μ-slip curve can be used to estimate the adhesion. The relationship between the road adhesion coefficient and the slope of its μ-slip curve can be described by the linear equation shown in Eq. (4.57), where $\xi_{\mu\kappa}$ and $\xi_{\mu 0}$ are constants calibrated from simulation or experimental data.

$$\mu = \xi_{\mu\kappa} \cdot \kappa_\mu + \xi_{\mu 0} \tag{4.57}$$

Therefore, κ_μ is the key parameter for observing the road adhesion coefficient. This book implements real-time identification of the parameter κ_μ based on the RLS parameter identification method. The parameter identification in the road friction coefficient observation algorithm is described in Eq. (4.58), where the system output $y_\mu(t)$ is the longitudinal tire force, $\varphi_\mu(t) = \lambda \cdot F_z$ is the system input, $\theta_\mu(t) = \kappa_\mu$ is the unknown parameter, and $e_\mu(t)$ is the identification error between the system output $y_\mu(t)$ and its observed value $\varphi_\mu^T(t) \cdot \theta_\mu(t)$.

$$y_\mu(t) = \varphi_\mu^T(t) \cdot \theta_\mu(t) + e_\mu(t) \tag{4.58}$$

The following are the iterative computation steps for the system parameter identification method for road friction coefficient observation based on RLS.

1) Calculate the system output $y_{\mu,k} = F_{x,k}$ and system input $\varphi_{\mu,k} = \lambda_k \cdot F_{z,k}$ in Eq. (4.58) based on the calculated slip ratio, vehicle vertical load, and longitudinal tire force.

2) Calculate the parameter identification error $e_{\mu,k}$, as presented in Eq. (4.59), which represents the difference between the actual system output at time $t = k$ and the predicted system output at time $t = k-1$.

$$e_{\mu,k} = y_{\mu,k} - \varphi_{\mu,k}^T \theta_{\mu,k-1} = F_{x,k} - \lambda_k \cdot F_{z,k} \cdot \kappa_{\mu,k-1} \tag{4.59}$$

3) Compute the update gain $K_{\mu,k}$ and covariance matrix $P_{\mu,k}$ of the parameter identification system:

$$K_{\mu,k} = \frac{P_{\mu,k-1} \cdot \lambda_k \cdot F_{z,k}}{\gamma_\mu + F_{z,k} \cdot \lambda_k \cdot P_{\mu,k-1} \cdot \lambda_k \cdot F_{z,k}} \tag{4.60}$$

$$\begin{aligned} P_{\mu,k} &= \frac{1}{\gamma_\mu} \left[P_{\mu,k-1} - \frac{P_{\mu,k-1} \cdot \lambda_k \cdot F_{z,k} \cdot F_{z,k} \cdot \lambda_k \cdot P_{\mu,k-1}}{\gamma_\mu + F_{z,k} \cdot \lambda_k \cdot P_{\mu,k-1} \cdot \lambda_k \cdot F_{z,k}} \right] \\ &= \frac{1}{\gamma_\mu} \left[P_{\mu,k-1} - \frac{P_{\mu,k-1} \cdot \lambda_k \cdot F_{z,k} \cdot F_{z,k} \cdot \lambda_k \cdot P_{\mu,k-1}}{\gamma_\mu + F_{z,k} \cdot \lambda_k \cdot P_{\mu,k-1} \cdot \lambda_k \cdot F_{z,k}} \right] \end{aligned} \tag{4.61}$$

4) Update the observed identification parameter:

$$\kappa_{\mu,k} = \kappa_{\mu,k-1} + K_{\mu,k} \cdot e_{\mu,k} \tag{4.62}$$

Thus, when the vehicle performs driving or braking maneuvers on a straight road, κ_μ can be observed from Eq. (4.62), and then road friction coefficient can be observed based on Eq. (4.57).

4.3.2 ESTIMATION OF ROAD FRICTION COEFFICIENT BASED ON VEHICLE DYNAMIC NONLINEARITY

When the vehicle performs steering maneuvers, a comprehensive algorithm based on the nonlinearity of vehicle dynamics is employed to observe the road friction coefficient. As shown in Fig. 4.4, we set the maximum road friction coefficient $\mu_{\max}$ and the minimum road friction coefficient $\mu_{\min}$ to constant values of 0.8 and 0.2, respectively. Additionally, $\mu_{\max}$ can be adjusted during the estimation process of road friction coefficient. The calibration parameters $\varphi_{f,1}$, $\varphi_{f,2}$, $a_{y,f,1}$, and $a_{y,f,2}$ are determined based on the simulation data and the real-car test data. The nonlinearity $\Delta\varphi$ and Δa_y can be calculated as follows:

$$\begin{cases} \Delta\varphi = \rho_{\mu,f}\dfrac{l_f + l_r}{v_x}\left|\varphi_{ref} - \varphi\right| \\ \Delta a_y = \rho_{\mu,f}\dfrac{l_f + l_r}{v_x^2}\left|a_{y,ref} - a_y\right| \end{cases} \tag{4.63}$$

where $a_{y,ref} = v_x \cdot \varphi_{ref}$, and φ_{ref} is the reference yaw rate, while $\rho_{\mu,f} = 1 + v_x^2/v_{x,ch}^2$.

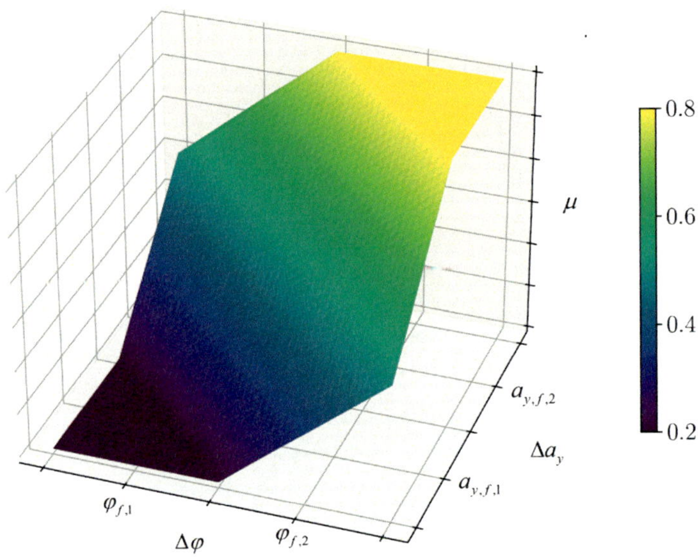

Figure 4.4 Schematic diagram of the relationship between road friction coefficient and vehicle dynamic nonlinearity indicator.

Furthermore, based on the nonlinearity $\Delta\varphi$ and Δa_y, we can yield $\hat{\mu}_\varphi$ and $\hat{\mu}_{ay}$, which are calculated as follows:

$$\hat{\mu}_\varphi = \begin{cases} \mu_{\max}, \Delta\varphi < \varphi_{f,1} \\ \mu_{\max} + \dfrac{\mu_{\max} - \mu_{\min}}{\varphi_{f,1} - \varphi_{f,2}}(\Delta\varphi - \varphi_{f,1}), \text{else} \\ \mu_{\min}, \Delta\varphi > \varphi_{f,2} \end{cases} \tag{4.64}$$

$$\hat{\mu}_{ay} = \begin{cases} \mu_{\max}, \Delta a_y < a_{y,f,1} \\ \mu_{\max} + \dfrac{\mu_{\max} - \mu_{\min}}{a_{y,f,1} - a_{y,f,2}}(\Delta a_y - a_{y,f,1}), \text{else} \\ \mu_{\min}, \Delta a_y > a_{y,f,2} \end{cases} \tag{4.65}$$

Finally, the estimated road friction coefficient in steering scenarios can be calculated as the weighted average of $\hat{\mu}_\varphi$ and $\hat{\mu}_{ay}$:

$$\hat{\mu} = \frac{w_{\mu,\varphi}\hat{\mu}_\varphi + w_{\mu,ay}\hat{\mu}_{ay}}{w_{\mu,\varphi} + w_{\mu,ay}} \tag{4.66}$$

4.4 ANALYSIS OF SIMULATION RESULTS

To verify the effectiveness of the proposed method for observing key chassis dynamic states (vehicle sideslip angle and road friction coefficient), this section presents the results analysis of simulations conducted in typical conditions.

4.4.1 SIMULATION OF PERFORMING SINUSOIDAL-STEERING MANEUVERS ON DRY ASPHALT ROADS

We first perform sinusoidal-steering maneuvers on a dry asphalt road to investigate the performance of our proposed method. The initial vehicle speed is set to 72 km/h, the steering amplitude is 90°, and the operation frequency is 4 Hz, as shown in Fig. 4.5 (a). The road friction coefficient of the dry asphalt road is 0.75. As shown in Fig. 4.5 (b) and (c), vehicle lateral acceleration and yaw rate are depicted. Both curves exhibit a similar pattern, characterized

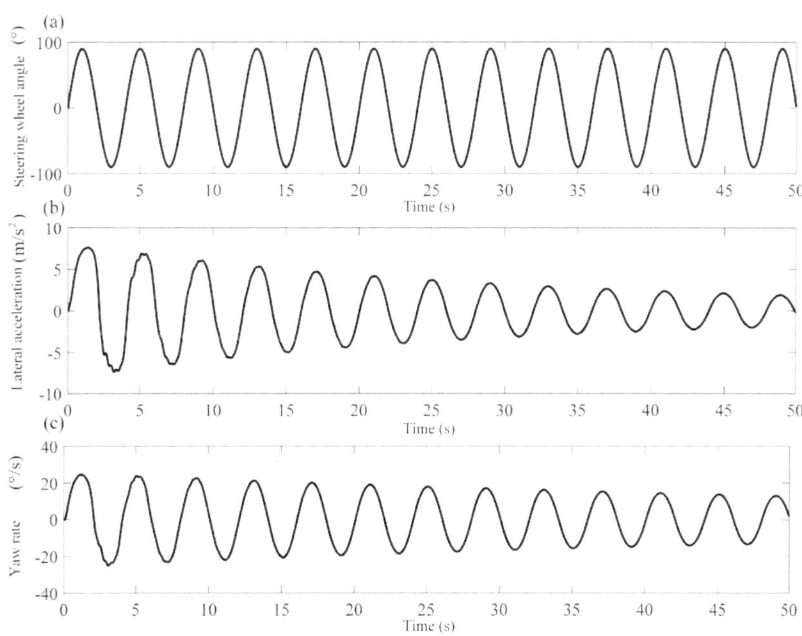

Figure 4.5 Steering wheel angle and vehicle dynamic states in this sinusoidal steering simulation test on dry asphalt roads, (a) steering wheel angle, (b) lateral acceleration, (c) vehicle yaw rate.

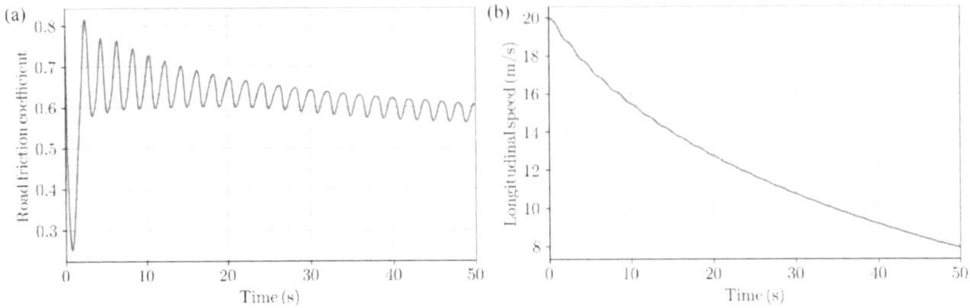

Figure 4.6 Road friction coefficient and vehicle speed in this sinusoidal-steering simulation test on dry asphalt roads, (a) road friction coefficient, (b) vehicle longitudinal speed.

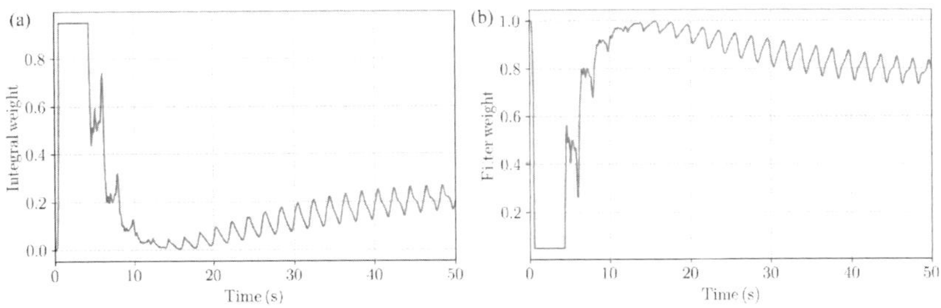

Figure 4.7 Fusion factors in this sinusoidal-steering simulation test on dry asphalt roads, (a) integral weight, (b) filter weight.

by a gradual sinusoidal variation corresponding to the steering wheel angle. The reduction in vehicle speed during coasting contributes to the decreasing amplitude of both curves. The peak lateral acceleration is approximately 7.6 m/s^2, and the peak yaw rate is approximately 24.6°/s.

Figs. 4.6, 4.7, and 4.8 present the observation results for the sinusoidal-steering simulation on a dry asphalt road. Fig. 4.6(a) represents the estimated road friction coefficient. It is evident that the friction coefficient fluctuates significantly at the beginning of the simulation but quickly converges to the road friction coefficient value of 0.75. It maintains proximity to this value throughout the simulation, gradually converging toward the actual road friction coefficient. Fig. 4.7 shows the adaptive fusion factors of the estimated vehicle sideslip angles from the ASCKF algorithm and the integration calculation.

Combining with Fig. 4.5 (b) and (c), it can be seen that in the initial 8 s of the simulation, the vehicle speed is relatively high, and vehicle dynamics exhibits nonlinear traits, resulting in large observation errors in the SCKF and ASCKF filter algorithms for vehicle sideslip angle, with SCKF showing the largest error and ASCKF providing a slight improvement. The adaptive fusion algorithm proposed in this chapter shows the smallest deviation from the actual value. During this period, the weight assigned to the integration calculation method is high, and the weight allocated to the filter observation is low.

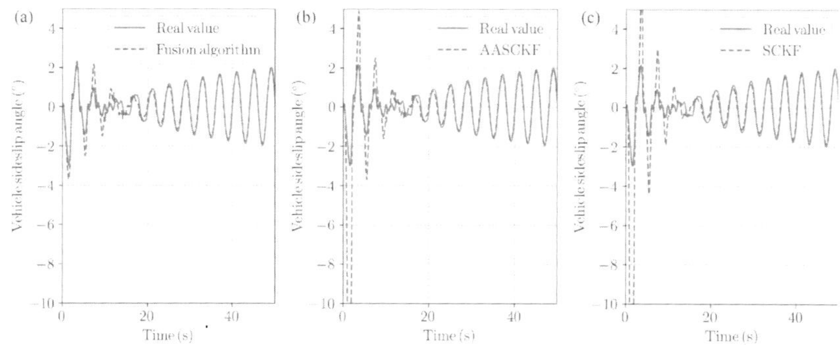

Figure 4.8 Simulation results of vehicle sideslip angle observation in this sinusoidal-steering simulation test on dry asphalt roads. (a) fusion algorithm, (b) AASCKF, (c) SCKF.

Table 4.1

Error analysis of simulation observations on dry asphalt roads.

Metric	SCKF	ASCKF	Adaptive fusion
RMSE	2.104 °	1.845 °	0.292°
Peak error	16.653 °	14.818 °	1.268°

Subsequently, as the simulation progresses, the weight for filter observation gradually increases with fluctuations. After 8 s, the estimation errors of all observation methods gradually decrease, and the performance discrepancies become insignificant.

Table 4.1 presents detailed results for the observation of vehicle sideslip angle. As shown in Fig. 4.8, around 1.6 s, the errors of the SCKF and ASCKF algorithms reach their maximums, at 16.7° and 14.8°, respectively. At approximately 5.5 s, the maximum deviation of the proposed adaptive fusion algorithm is only 1.268°. The RMSE values achieved by the SCKF, ASCKF, and adaptive fusion algorithms are 2.104°, 1.845°, and 0.292°, respectively. These results highlight that in this sinusoidal-steering simulation on dry asphalt roads, the SCKF exhibits the largest error, with ASCKF slightly enhancing observation accuracy over SCKF, while the adaptive fusion algorithm achieves the highest accuracy.

4.4.2 SINUSOIDAL-STEERING SIMULATION ON LOW-FRICTION ROADS

We also conduct the sinusoidal-steering simulation on low-friction roads. The setup is as follows: the initial vehicle speed is 65 km/h, with sinusoidal-steering

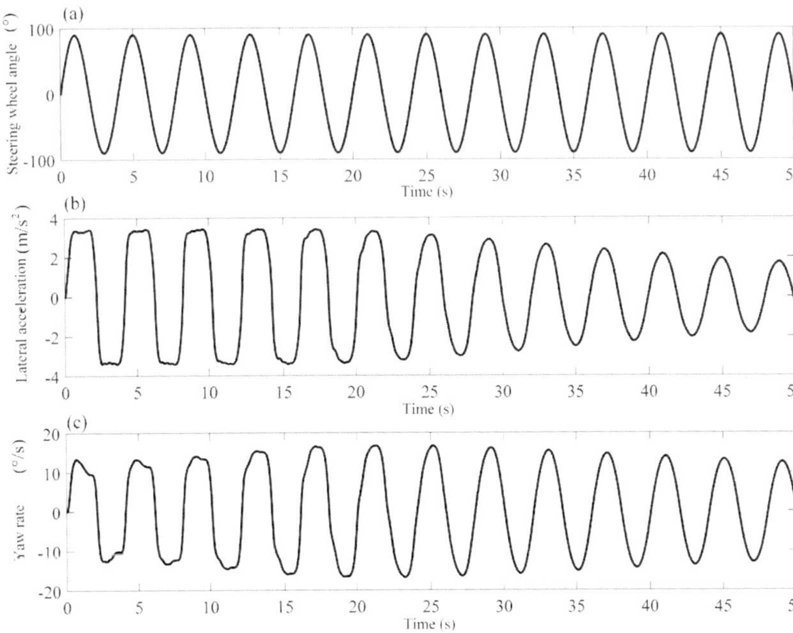

Figure 4.9 Steering wheel angle and vehicle dynamic states in the sinusoidal-steering simulation test on low-friction roads.

operations similar to the simulation of performing sinusoidal-steering maneuvers on dry asphalt roads. The steering wheel angle amplitude is 90°, and the steering operation frequency is 4 Hz, with a road friction coefficient of 0.35. Fig. 4.9 presents vehicle states in this low-friction scenario: vehicle lateral acceleration and yaw rate change in a quasi-sine pattern with the steering wheel angle. The peak lateral acceleration in this simulation is approximately 3.4 m/s^2, and the peak yaw rate is about 16.7°/s.

Figs. 4.10, 4.11, and 4.12 present the observation results for road friction coefficient and vehicle sideslip angle in the low-friction simulation. Fig. 4.10(a) depicts the observation results of road friction coefficient. In the initial 3 s of the simulation, the road friction coefficient fluctuates significantly before quickly converging to 0.35. Subsequently, the estimated value fluctuates around this value, closely approximating the actual road friction coefficient. Fig. 4.11 illustrates the adaptive fusion factor variation curve of the adaptive fusion algorithm in this low-friction test. In this scenario, the vehicle is prone to exhibit nonlinear dynamics, particularly when vehicle lateral acceleration and yaw rate are relatively high, which enhances the presence of nonlinear dynamic characteristics. As a result, the weights of filter observation and integration calculation exhibit a similar sinusoidal pattern with the steering wheel angle, demonstrating significant fluctuations.

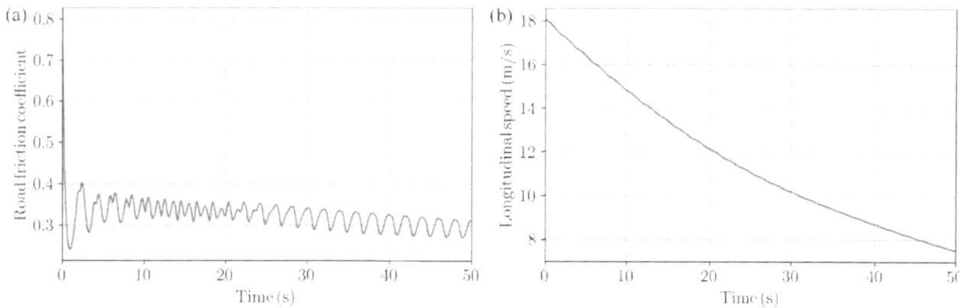

Figure 4.10 Road friction coefficient and vehicle speed curves in the low-friction simulation, (a) road friction coefficient, (b) vehicle speed.

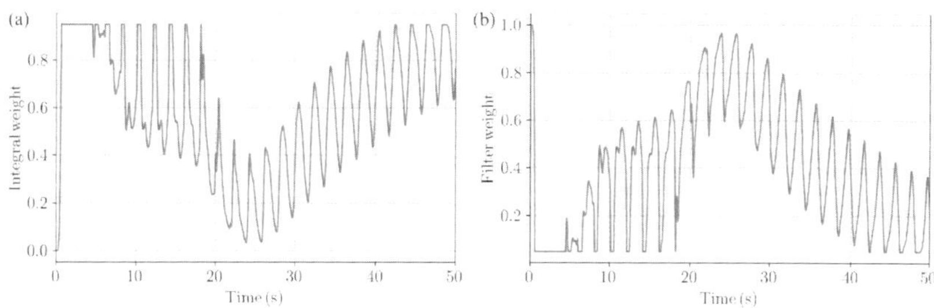

Figure 4.11 Fusion factors in the low-friction simulation, (a) integral weight, (b) filter weight.

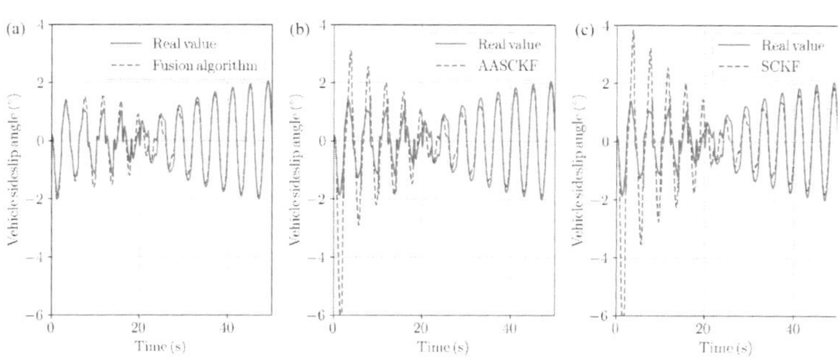

Figure 4.12 Observation results of vehicle sideslip angle in the low-friction simulation, (a) fusion algorithm, (b) AASCKF, (c) SCKF.

Fig. 4.12 and Table 4.2 present detailed results for vehicle sideslip angle observation in the low-friction simulation. For approximately the initial 25 s of the low-friction simulation, the performance differences among the different observation algorithms are significant: the SCKF algorithm exhibits the largest error, the ASCKF algorithm shows a slight improvement in observation accuracy compared to SCKF, and the adaptive fusion algorithm proposed in this chapter achieves the highest accuracy, with the observed curve

Table 4.2
Error analysis of simulation observations on low-friction road.

Metric	SCKF	ASCKF	Adaptive fusion
RMSE	0.960°	0.782°	0.228°
Peak error	5.795°	4.960°	0.665°

closely matching the curve of actual vehicle sideslip angles. After 25 s, the performance differences among the three methods gradually diminish, with all methods achieving high accuracy in observing vehicle sideslip angle. Specifically, in the low-friction simulation, the maximum errors achieved by SCKF, ASCKF, and the adaptive fusion algorithm are 5.795°, 4.960°, and 0.665°, respectively. Additionally, the RMSE values are 0.960°, 0.782°, and 0.228°, respectively. This demonstrates that in the low-friction simulation, the adaptive fusion algorithm proposed in this chapter has a significant advantage over both SCKF and ASCKF algorithms.

4.5 EXPERIMENTAL VALIDATION

This section further presents an analysis of experimental data on vehicle sideslip angle and road friction coefficient observation to validate the effectiveness of the proposed method in this chapter.

4.5.1 DUAL-ANTENNA DIFFERENTIAL GPS MEASUREMENT OF VEHICLE SIDESLIP ANGLE

To measure vehicle sideslip angle in real-time during the experiment, we employ a dual-antenna differential GPS (From NovAtel corporation) with a positional accuracy of 0.02 m. As shown in Fig. 4.13, two mobile stations, A

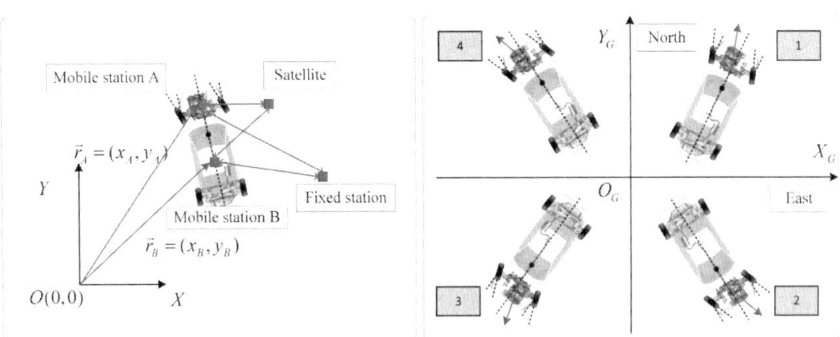

Figure 4.13 Schematic diagram of differential GPS measurement principle.

and B, are installed along the longitudinal axis of the experimental vehicle. The positions of A and B are received at fixed sampling intervals based on the communication with satellite S. The satellite communication with the mobile stations A and B is corrected via the fixed base station F to improve the differential GPS positioning accuracy to 0.02 m. For example, considering the vector direction of vehicle speed on the left side of the line connecting mobile stations A and B, the vehicle's attitude is divided into four quadrants. The sideslip angle is then calculated as shown in Eq. (4.67), where i_q is the quadrant number based on the vehicle's attitude. When $i_q = 1$ or $i_q = 2$, $j_q = 1$; when $i_q = 3$ or $i_q = 4$, $j_q = 2$. ϕ_v is the angle between the vector direction of vehicle speed and the north of the earth, which can be directly measured by the dual-antenna differential GPS, and ϕ_p is vehicle heading angle, calculated as shown in Eq. (4.68). When the vector direction of vehicle speed is on the right side of the line connecting mobile stations A and B, vehicle sideslip angle calculated by the above method is the negative of the actual value.

$$\beta = \frac{2+(-1)^{j_q}}{2} \cdot \pi + (-1)^{i_q} \cdot \phi_p - \phi_v \tag{4.67}$$

$$\phi_p = \arctan |y_A - y_B| / |x_A - x_B| \tag{4.68}$$

4.5.2 ANALYSIS OF EXPERIMENTAL DATA

Fig. 4.14(a) illustrates the driving maneuvers in the real-car experiment. The initial vehicle speed is 40 km/h. In the initial 8 s of the experiment, the vehicle accelerates to approximately 60 km/h, coasts for a while, and then decelerates slowly to 35 km/h. At about 10 s, the vehicle performs a sinusoidal-steering operation with gradually increasing amplitude, with a period of about 3.5 s. The peak value of the steering wheel angle achieves about 300° at about 31 s. Fig. 4.14(b) and Fig. 4.14(c) illustrate the changes in vehicle lateral acceleration and yaw rate in the experiment. The estimated road friction coefficient

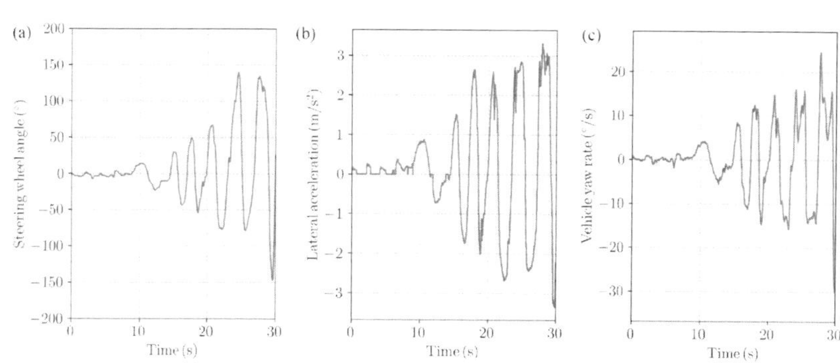

Figure 4.14 Steering wheel angle and vehicle dynamic states in the real-car experiment, (a) steering wheel angle, (b) lateral acceleration, (c) vehicle yaw rate.

quickly converges and remains around 0.2 before 10 s, then fluctuates between 0.2 and 0.4 thereafter, as demonstrated in Fig. 4.15.

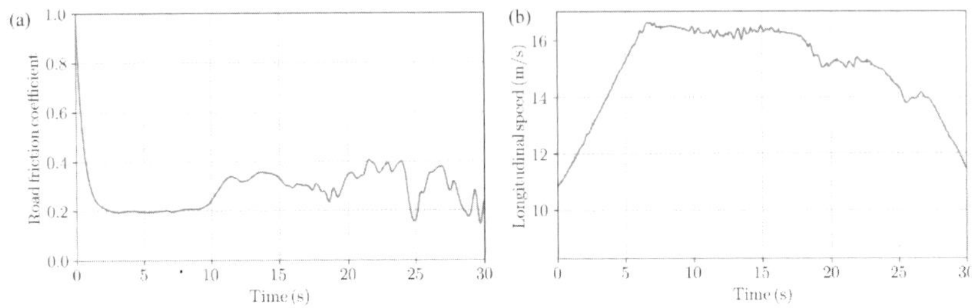

Figure 4.15 Road friction coefficient and vehicle speed in the real-car experiment, (a) road friction coefficient, (b) vehicle speed.

Figs. 4.16 and 4.17 show the observation results of the algorithms in the experiment. The curves of the adaptive fusion factors indicate that before approximately 18 s in the real-vehicle experiment, the filtering observation weight was significantly higher, while the integration calculation weight was relatively low. Subsequently, the weights fluctuated with the vehicle dynamic states. The observation results of vehicle sideslip angle are shown in Fig. 4.17. During the initial 18 s of the experiment, the observations of all three methods closely match the actual values. Afterward, the errors in the SCKF and ASCKF algorithms increase, while the adaptive fusion algorithm's observation curve shows the smallest deviation from the curve of the measurement, with an RMSE value of 1.064°. It is clear from Fig. 4.17 that the observation performance of the adaptive fusion algorithm is superior to both SCKF and ASCKF algorithms. However, the error of the adaptive fusion algorithm in the real-vehicle experiment is slightly larger than that in the simulation. This discrepancy is caused by the influence of road gradient and slope, as well as the lateral motion of the vehicle's sprung mass, which affects the sensor measurements.

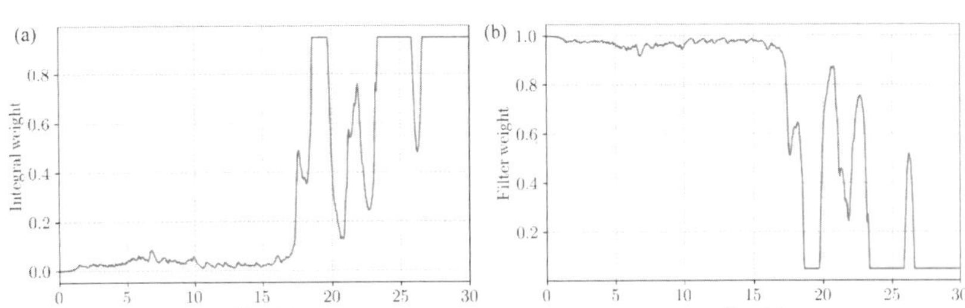

Figure 4.16 Curves of the fusion factors in the real vehicle experiment, (a) integral weight, (b) filter weight.

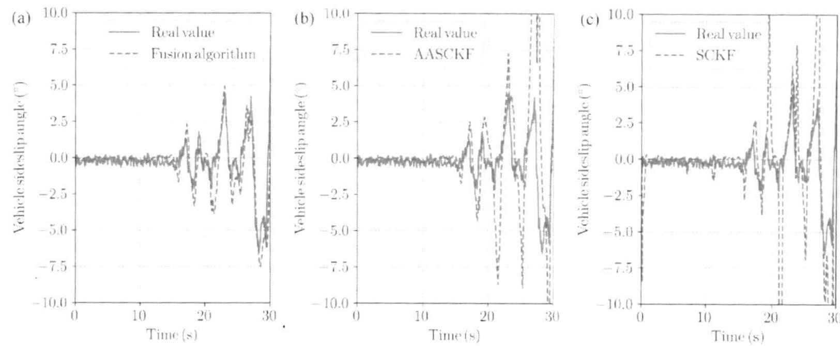

Figure 4.17 Observation results of vehicle sideslip angle in the real-vehicle experiment, (a) fusion algorithm, (b) AASCKF, (c) SCKF.

4.6 CONCLUSION

This chapter investigates methods for estimating key dynamic states (vehicle sideslip angle and road friction coefficient), and validates the proposed estimation methods through simulation analysis and real-car experiments. The nonlinearity of vehicle dynamics in complex driving conditions poses significant challenges to accurately estimating dynamic states such as the vehicle sideslip angle. To address this issue, two observation methods are designed: the zero-reset integration method based on sensor signals and the ACKF method. An adaptive fusion algorithm is then employed to integrate the estimated values from these two methods according to vehicle dynamic states. This approach enabled accurate estimation of the vehicle sideslip angle through adaptive fusion. To address the challenge of estimating road friction coefficient in complex driving conditions, this chapter presents the design of a road friction coefficient mode-switching observer. This observer combines an RLS identification method with a compensation algorithm, enabling a comprehensive estimation of road friction coefficient. Simulation and real-vehicle experiments in various conditions are conducted to validate the effectiveness of the proposed methods. The results demonstrate that the adaptive fusion observation algorithm developed in this chapter can accurately estimate vehicle sideslip angles in diverse and complex driving scenarios, and the estimated road friction coefficient values meet performance requirements. In dry asphalt and low-friction road simulations, the RMSE values of the adaptive fusion algorithm are 0.292° and 0.228°, respectively, while the real-vehicle test error was 1.064°. The accuracy of the adaptive fusion algorithm outperforms the SCKF and ASCKF methods.

5 Chassis-Domain-Aware Motion Planning

This chapter focuses on the critical challenge of performing safe and feasible trajectory planning for AVs in highly complex and uncertain driving environments. The trajectory planning module generates collision-free paths; however, the high nonlinearity of vehicle dynamics might lead to handling instability when tracking dynamically infeasible paths. This chapter introduces a quantitative assessment method for evaluating AVs' chassis-domain dynamic performance. A rigorously defined metric is derived, which enables a comprehensive evaluation of safety and comfort performance. Based on this foundation, this chapter develops a chassis global dynamics-oriented motion planning approach. Optimal trajectories are produced via numerical optimization, and the trajectories satisfy geometric smoothness, collision-free safety, and dynamic feasibility. The proposed evaluation methodology and motion-planning approach are validated through extensive tests across various scenarios. Results illustrate the effectiveness of the proposed methods. This chapter provides essential insights into the dynamics-based performance appraisal and motion planning of AVs.

5.1 BACKGROUND

Existing studies on motion planning are broadly categorized into global trajectory planning and local trajectory planning [122, 32]. The former focuses on determining feasible routes from the start to the destination, prioritizing some performance aspects such as shortest time, minimal distance, or lowest cost [149]. On the other hand, the latter is responsible for outputting feasible, collision-free trajectories based on global routes and real-time environments. The local trajectories are utilized as the desired paths for chassis systems to follow. Therefore, the motion planning module must adapt to the uncertain surroundings and the nonlinear vehicle dynamics, ensuring safety and feasibility, which remains a challenging issue. This chapter is dedicated to analyzing aspects of vehicle dynamic performance and local motion planning and sharing new insights into chassis-domain-aware motion planning for AVs. First, this section will review a variety of investigations on vehicle dynamics performance and local motion planning. Some fundamental knowledge and definitions will also be discussed.

DOI: 10.1201/9781003583929-5

5.1.1 EVALUATION OF VEHICLE DYNAMICS PERFORMANCE

Vehicle dynamics has been an important focus in automotive engineering over the past few decades. Earlier investigations primarily concentrated on enhancing specific aspects of chassis dynamic performance by optimizing the design of some chassis components or developing individual chassis control subsystems. For example, the electronic stability control system enforces differential braking operations to improve chassis dynamic stability, while the active suspension system can improve vehicle performance of riding comfort via actively intervening in suspension components. The two above aspects of chassis dynamic performance are essential to performing autonomous driving, so this subsection will discuss them in detail.

For chassis dynamic stability, many researchers have made significant efforts to study related key issues, especially the improvement of chassis handling performance. At the same time, the assessment of chassis dynamic stability is the foundation for optimizing the design of chassis components and developing active control strategies [45, 74]. Several classical methods have been studied extensively to evaluate chassis handling stability as shown in Fig 5.1. Vehicle stability control systems usually utilize the dynamic response deviation approach to assess vehicle steering performance, and their control logic is based on measuring the deviation values of vehicle yaw rate and sideslip angle [26, 82]. The phase portrait method could provide a more accessible way to visualize vehicle stability regions. When plotting vehicle yaw rate and sideslip angle on a phase plane, the stability status can be examined by observing whether the trajectories could finally return to the equilibrium point [66, 121]. Moreover, vehicle stability conditions can be derived by treating the vehicle chassis system as a control system and using mathematical control theories, such as the Lyapunov stability analysis technique and Hurwitz criteria for checking stability [144, 146]. Despite these developments, dynamic stability analysis remains insufficient in the context of autonomous driving. The aforementioned stability regions or criteria might fail to capture the full complexity of vehicle dynamics, especially when considering the interactions of vehicle longitudinal, lateral, and vertical movements.

Maximizing passenger ease and satisfaction is the main goal when it comes to riding comfort. A vehicle's suspension system is essential for reducing vehicle vibrations caused by uneven road surfaces, and significant advancements have been made in active suspension control strategies [102, 143]. However, riding comfort levels, which are influenced by subjective perceptions, differ among different passengers. As a result, quantitatively evaluating riding comfort level is crucial but remains an opening challenge. Some studies have used survey data to investigate how driving behaviors vary across different drivers [97]. Objective data, such as acceleration and jerk, have been analyzed to assess different driving styles [36, 11], and vehicle jerk was used to evaluate riding comfort in the literature [92]. In the study [39], a neural network-based augmentation approach was introduced to evaluate riding comfort, and data

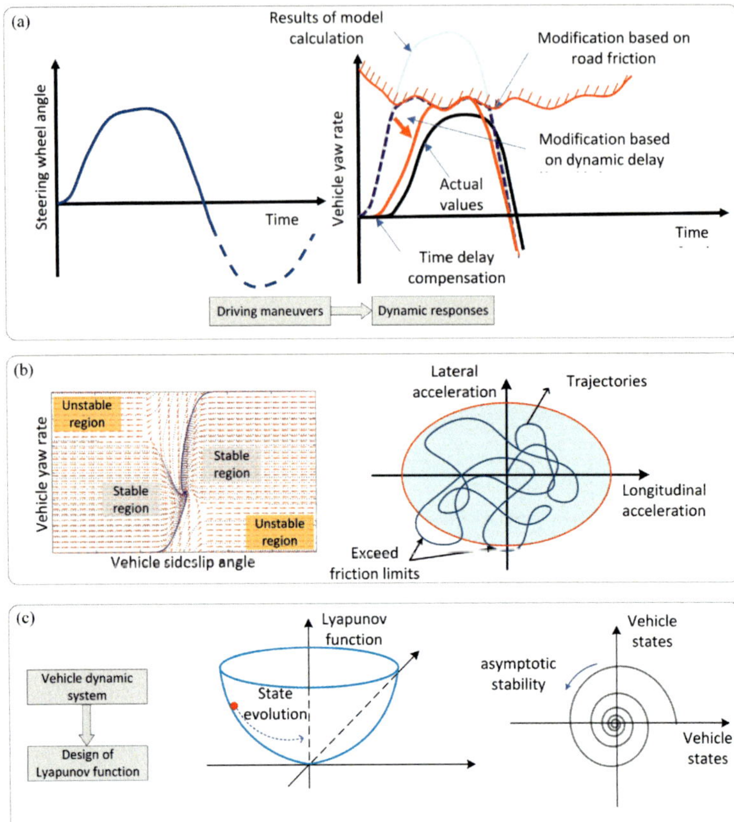

Figure 5.1 Classical approaches of evaluating chassis dynamic stability, (a) dynamic response deviation; (b) phase portrait; (c) stability analysis based on the control system theory.

segmentation and resampling were used in this method. Principal component analysis and K-means clustering were investigated to design a driving style categorization method [35].

5.1.2 MOTION PLANNING METHODS

Considerable work on motion planning has been reported in recent years. This chapter divides existing local trajectory planning methods into five types, i.e., interpolating curve-based planners, sampling-based planners, graph search-based planners, artificial potential field-based planners, and numerical optimization-based planners [47]. Details can be found in Table. 5.1 [131]. Artificial potential field-based planners and numerical optimization-based planners are extensively used among them. For example, the potential field method was used to design a model predictive path planning [112]. Besides, Several

Table 5.1
Comparison of different motion planning methods.

Motion planning methods	Technique	Description
Interpolating curve-based planners	Lines and circles, polynomial curves, clothoid curves, etc.	This method uses predefined reference points to generate a smoother path, considering vehicle constraints and dynamic traffic scenarios.
Sampling-based planners	Rapidly-exploring random tree, probabilistic roadmap method, etc.	This method performs random sampling of the state space to find connectivity within it.
Graph search based planners	A-star algorithm, Dijkstra algorithm, etc.	This method usually utilizes an occupancy grid or lattice to represent the state space and uses graph search algorithms to generate a feasible path.
Artificial potential field-based planners	Virtual force field, velocity vector field, etc.	This method utilizes specialized potential functions to represent various traffic elements, for example, road geometry, and obstacles, and generates optimal paths by determining the descent direction of the potential field.
Numerical optimization-based planners	Road constraints, ego-vehicle constraints, etc.	This method usually refines precalculated trajectories through optimizing a defined cost function subject to a variety of constraints.

comprehensive motion planning frameworks were proposed based on the potential field modeling [56, 61]. Some investigations based on numerical optimization methods were reported [17], and some important variables such as vehicle acceleration, jerk, and speed were involved to create optimal paths.

The Cartesian coordinate and Frenet coordinate are the two coordinate systems usually used in existing trajectory planning studies [88]. Autonomous vehicles usually rely on the Cartesian coordinate, defined by the x and y axes, for navigation and localization. However, planning paths directly in Cartesian coordinates is challenging in the case of curved roads. To address this issue, some studies have introduced the Frenet coordinate, which can simplify trajectory planning by transforming irregularly shaped real-world roadways into linear representations within the Frenet frame [139, 84]. Concretely, some investigations which were conducted in the Cartesian frame were reported, for example, a model predictive method of generating kinematically feasible paths in the Cartesian frame [85], and an optimal trajectory planning method within the Cartesian frame in the scenario of curvy roads [85]. The widely adopted Frenet frame is suitable to deal with the scenarios of curvy roads because it could transform curvy roads into straight "tunnel-like" representations with left and right boundaries [139]. An iterative optimization method using the Frenet frame was introduced by the literature [141]. The combination of both coordinate systems was also investigated to perform local path planning. In the literature [84], the Cartesian frame was utilized to determine optimal paths through a designed objective function, and the Frenet frame was adopted to decouple the path planning into spatial path planning and velocity planning. In the literature [60], a set of candidate paths was generated and subsequently transformed into the Cartesian frame so that an optimal trajectory that prioritized both safety and passenger comfort could be determined.

5.1.3 PROBLEM STATEMENT

The tremendous promise of self-driving cars is dependent on their ability to safely perform driving maneuvers in dynamic, complicated traffic conditions. The path planning module is a crucial component, responsible for generating optimal local paths. Furthermore, the desired paths are tracked by the implementation of various chassis subsystems. Existing path planning approaches mainly focus on the issue of geometric safety, including collision avoidance, feasibility of path geometry, etc. Most prevailing works proposed various approaches for the performance improvement of navigation and geometric interaction in dynamic traffic environments and surroundings. However, we claim that existing methods are still insufficient because the performance and impact of vehicle chassis dynamics are not considered.

As we all know, the driving performance of an autonomous vehicle is determined ultimately by chassis dynamics, which is performed by chassis driving, steering, and braking subsystems. Vehicle chassis dynamics is highly nonlinear when a car operates near its handling limits, which might lead to dynamic instability or even rapid spinning of the vehicle body. If the desired path generated by the path planning module is dynamically infeasible, the vehicle may lose its dynamic stability when performing path-tracking maneuvers due to the high nonlinearity of vehicle dynamics. Therefore, we argue that the performance of chassis dynamics must be considered in local path planning modules, which remains unsolved. This chapter introduces a chassis-global-dynamics-oriented path planning solution. This approach integrates the performance of chassis dynamics into the path planning procedure so that the seamless alignment between the desired trajectories and chassis executions can be guaranteed.

5.2 COMPREHENSIVE ASSESSMENT OF CHASSIS DOMAIN DYNAMICS

The mass deployment of autonomous vehicles necessitates that an autonomous vehicle can perform with experienced driving skills to ensure overall performance. Hence, efficient testing and evaluation of autonomous vehicles' functionality and comprehensive performance are essential. Vehicle performance consists of various factors, including safety, driving comfort, and fuel efficiency, so achieving a comprehensive evaluation is challenging. Since autonomous vehicles are safety-critical systems, various aspects of safety have been investigated for decades. However, existing research on performance evaluation mainly emphasizes driving safety considering traffic participants, particularly in the context of collision avoidance with surroundings. The performance assessment of chassis dynamics, for example, dynamic stability and riding comfort, remains unsolved.

As illustrated in Fig. 5.2, this section introduces a quantitative assessment method for integrated dynamic performance of vehicle chassis domain. Chassis spatio-temporal states are analyzed to figure out their variations and correlations. Then, the stable boundaries of chassis dynamics are derived, which provides a benchmark for evaluating vehicle internal dynamic safety. A rigorous unified metric for chassis domain dynamics is designed to assess integrated dynamic stability and riding comfort performance, which can reflect vehicle longitudinal, lateral, and vertical motions. This unified metric provides a quantitative manner to evaluate chassis dynamic performance and can be directly used in the numerical optimization procedure of the motion planning task.

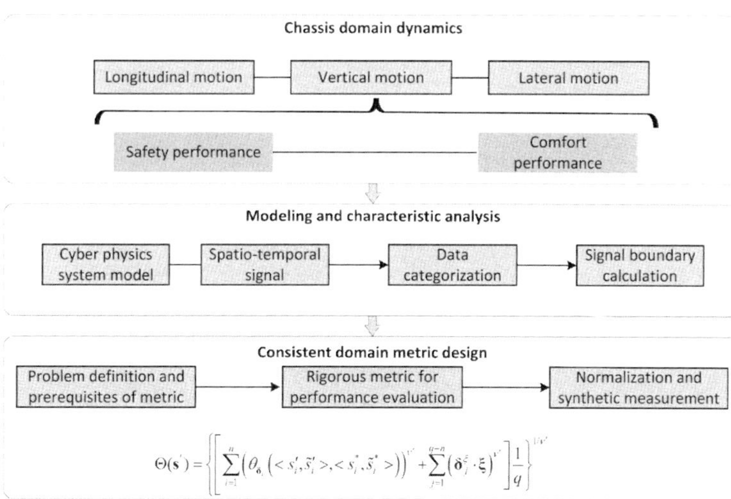

Figure 5.2 Illustration of chassis domain performance evaluation.

5.2.1 ANALYSIS OF CHASSIS DYNAMICS

This part begins with an extensive analysis of the vehicle chassis system, which consists of the modeling of nonlinear dynamics and vehicle spatio-temporal signals. Vehicle longitudinal dynamics, lateral dynamics, and vertical dynamics have been introduced in Section 2.4 and Section 3.2, so this chapter will not go into details. While this chapter has to discuss further the load transfer ratio (LTR) with respect to vehicle rolling motion. The LTR, which is defined as shown in Eq. (5.1), quantifies the degree of vehicle vertical load variation i.e., vertical tire force F_z enforced on the left and right sides.

$$\text{LTR} = \left[\left(F_{zfl} + F_{zrl}\right) - \left(F_{zfr} + F_{zrr}\right)\right] / mg \tag{5.1}$$

Some dynamic states such as vehicle pitch and roll rates, are difficult to be measured in mass-produced vehicles, so the first step is to analyze and identify key signals that can be directly measured, enabling the effective implementation of the proposed method in real cars. This subsection will investigate the statistical features and boundary conditions of various vehicle dynamic states. The variation tendencies of different vehicle dynamic signals are evaluated in typical steering scenarios, and the experimental scenarios involve accelerating, braking, and steering operations. Fig. 5.3 presents their cross-correlation analysis, and it reveals that chassis dynamic states representing vehicle motion in the same direction have strong cross-correlations. These findings align with the principles of vehicle dynamics, for example, vehicle longitudinal acceleration and pitch rate have correlation values of 0.54 and −0.34 with vehicle longitudinal slip ratio, respectively, and obvious correlations can also be observed in vehicle lateral and roll motion dynamics, as presented in Fig. 5.3.

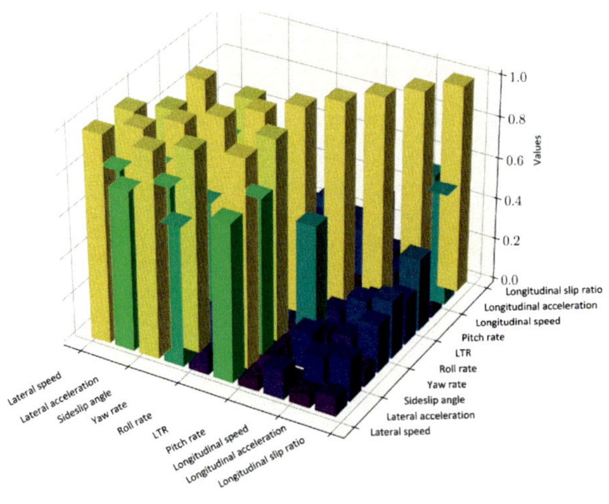

Figure 5.3 Analysis of relationships among vehicle spatio-temporal signals.

To identify the key factors impacting chassis domain performance, this chapter extracts vehicle spatio-temporal signals and categorizes them into three groups based on vehicle longitudinal, lateral, and vertical motions. First, the chassis actuators are required to guarantee chassis dynamic stability in the three directions of vehicle motions. Second, the passengers' vestibular systems do not directly sense specific dynamic states, and they detect changes in these states over time, i.e., their time derivatives. This plays a central role in riding comfort. As a result, this chapter pairs relevant performance-related signals and their derivatives, which can then be utilized to develop an effective evaluation metric, as demonstrated in Fig. 5.4. To normalize selected signals, we should then analyze the maximum values of key spatio-temporal signals. For some variables, such as the longitudinal slip ratio λ and LTR, this chapter defines their peak values as one based on their definitions. Moreover, the maximum value of the vehicle roll rate coincides with the maximum value of the LTR. Notably, the road friction coefficient μ, which defines the upper limit for tire forces, primarily influences the peak values of vehicle sideslip angle, acceleration, and yaw rate.

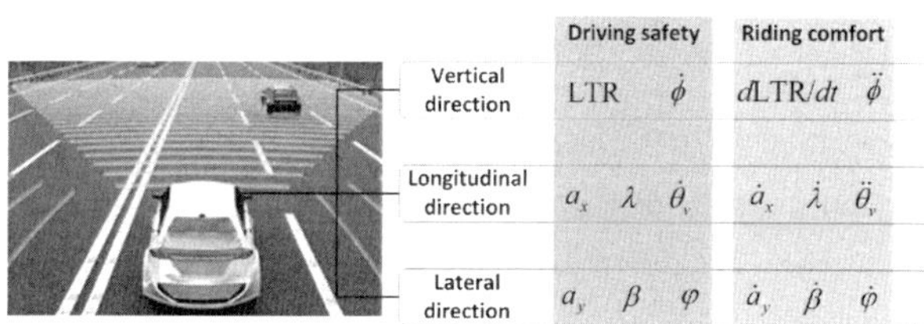

Figure 5.4 Classification of vehicle dynamic signals.

As discussed in Fig. 5.1, the G-G diagram, $\beta-\varphi$ phase portrait, and the derivation of stability conditions are commonly utilized methods for illustrating vehicle dynamic performance, and they are adopted in this chapter as well. Fig. 5.5 (a) depicts the G-G diagram based on some typical test data, demonstrating the trajectories of the longitudinal and lateral acceleration. The $\beta-\varphi$ phase portrait is presented in Fig. 5.5 (b).

Let us analyze the derivation of dynamic stability conditions to reveal the vehicle stable operating regions. According to vehicle bicycle dynamic model, this chapter uses Eq. (5.2) to describe the characteristics of vehicle sideslip angle and yaw rate [70]. Thus, we can yield a geometric representation of chassis dynamic stability through assessing whether the trajectory of vehicle sideslip angle and yaw rate returns to the equilibrium, which forms the phase portrait method [63].

$$\dot{\beta}=\frac{F_{yf}+F_{yr}}{mv_x}-\varphi,\quad \dot{\varphi}=\frac{l_fF_{yf}-l_rF_{yr}}{I_z} \tag{5.2}$$

This chapter analyzes the equilibrium of tire forces at their peak values, which implies $l_f\cdot F_{yf_{eq}}=l_r\cdot F_{yr_{eq}}$, based on which, the constraints of vehicle

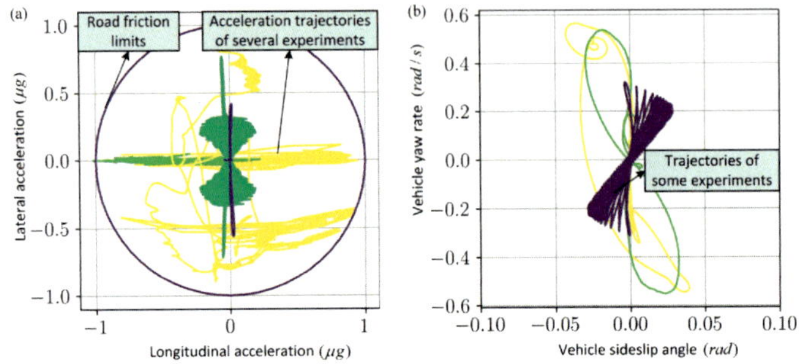

Figure 5.5 Trajectories of key dynamic states in several test scenes: (a) G-G diagram, (b) trajectories of $\beta - \varphi$.

yaw rate can be yielded. Using the bicycle dynamic model, the expression of φ_{eq} can be calculated in terms of either the front or rear tire forces as follows.

$$\frac{F_{yr_{eq}}(1+l_r/l_f)}{mv_x} - \frac{F_{yf_{eq}}(1+l_f/l_r)}{mv_x} \tag{5.3}$$

The constraints of vehicle yaw rate have to be calculated according to the lateral tire force enforced at the axle which reaches its saturation, which is presented as follows.

$$\varphi_{\max} = \begin{cases} \dfrac{F_{yr\max}(1+l_r/l_f)}{mv_x} & F_{yf_{\max}} \geq \dfrac{l_r}{l_f}F_{yr_{\max}} \\ \dfrac{F_{yf_{\max}(1+l_f/l_r)}}{mv_x} & F_{yf_{\max}} < \dfrac{l_r}{l_f}F_{yr_{\max}} \end{cases} \tag{5.4}$$

Considering that a car is required to retain sufficient front tire grip to guarantee the steering capability of the front wheel, this chapter derives the maximum permissible sideslip angle by analyzing the scenario where the sideslip angle of the rear tire reaches its saturation limit. Thus, the dynamic stability boundaries for both vehicle yaw rate and sideslip angle can be explicitly defined as follows.

$$|\varphi| \leq \frac{\mu F_{zr} + l_r \mu F_{zr}/l_f}{mv_x} = \frac{\mu g}{v_x} \tag{5.5}$$

$$\left| \beta - \frac{l_r \varphi}{v_x} \right| \leq \alpha_{r,\max} \tag{5.6}$$

where $\alpha_{r,\max}$ represents the maximum rear sideslip angle, which can be yielded based on the nonlinear brush tire model corresponding to peak lateral tire force.

As demonstrated in Fig. 5.6, this chapter conducts a comprehensive analysis using vehicle bicycle dynamic model and the nonlinear brush tire model [104] to determine the dynamic stability boundary. It can be found that the stable

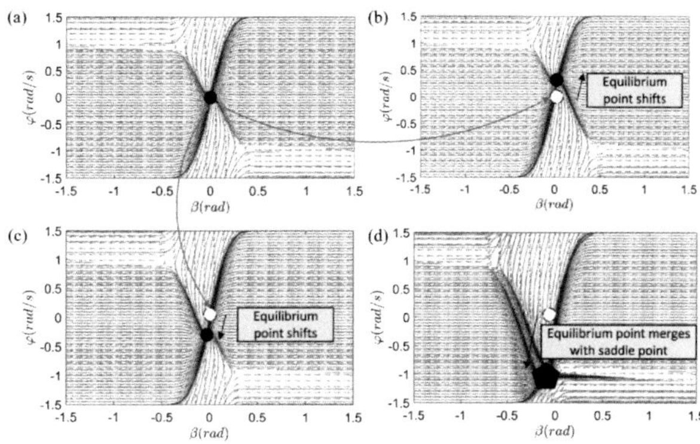

Figure 5.6 Demonstration of the $\beta - \varphi$ phase portrait derived from the vehicle bicycle dynamic model. (a) $v_x = 10\ m/s,\ \delta = 0^\circ$. (b) $v_x = 1\ m/s,\ \delta = 5^\circ$. (c) $v_x = 10\ m/s,\ \delta = -5^\circ$. (d) $v_x = 10\ m/s,\ \delta = -20^\circ$.

equilibrium point exists, and its position varies with the value of the steering angle input. In the scenario of $\delta = 0$, the equilibrium point coincides with the origin of the coordinate system. However, as δ increases, the equilibrium point shifts outward toward a saddle point corresponding to the steering direction, which lies along the curve of the peak stable yaw rate. The stable region is localized around the vicinity of the stable equilibrium point, representing a dynamically stable operating zone. It explicitly excludes the regions surrounding saddle points, since these regions are unstable and have the potential loss of vehicle control.

5.2.2 PRELIMINARY OF CONSISTENT METRIC

This subsection introduces the formulation of a general concept for the comprehensive evaluation of multi-performance criteria in complex systems involving various dynamic states. The comprehensive metric is required to quantify the "difference" between two sets of states of the dynamical system, which should be both mathematically consistent and physically interpretable.

First, we take a general multi-degree-of-freedom system Ψ as an example and consider its dynamic evolution. The system is described as $\dot{\mathbf{s}} = f(\mathbf{s}, \mathbf{u}, t)$, where $\mathbf{s} = [s_1, s_2, \ldots, s_n]$ denotes the dynamic state vector consisting of n states, $\mathbf{u}$ denotes the system input, and $f(\cdot)$ is the system function. Then, we define that $P_\Psi = p_1, p_2, \ldots, p_m$ represents the set of all performance indexes and there is a mapping relation $\rho : \mathbf{s}, t \to P_\Psi$ between the performance indexes and the system states.

Thus, this chapter aims to design a quantitative metric $\Theta : \mathbf{s}, t \times \mathbf{s}, t \to \mathbf{R}_+ = [0, +\infty)$, which can comprehensively evaluate each performance indicator $p_i \in P_\Psi$. This chapter will discuss the fundamental properties which are necessary for a robust metric such that consistency and applicability across various scenarios can be guaranteed. Furthermore, a specific metric that satisfies these criteria is instantiated and discussed in detail.

Specifically, The performance metric in this chapter is expected to quantify the distance between two sets of dynamic states effectively: the set of expected ground truth states $\mathbf{G} = g_1, g_2, \ldots, g_q$ and the set of measured states $\mathrm{s} = [s_1, s_2, \ldots, s_n]$. To ensure practical applicability and interpretable evaluation, a comprehensive metric must satisfy the following essential requirements:

1. **Capture Differences Effectively:** To achieve a clear divergence measure, the metric must accurately quantify the "difference" or distance between the two-state sets.
2. **Natural Physical Interpretation:** The metric should be consistent with physical phenomena so that it is intuitively understandable and interpretable in vehicle chassis domain dynamics.
3. **Transitivity Property:** The metric is required to meet the transitive property such it can keep consistency and ensure logical coherence in comparisons.
4. **Low Computational Complexity:** It is necessary to use minimal computational cost so that this metric could perform real-time implementation and has scalability for complex systems.
5. **Applicability to Finite Sets:** The metric must operate effectively on finite sets of states, hence its practical application in real-world scenarios can be ensured.

These requirements ensure that the proposed performance metric is both mathematically sound and practically viable for evaluating the dynamic performance of chassis domain dynamics.

Based on the above discussion, we can yield the following fundamental properties for all s^-, s', $\mathrm{s}^* \in \{\mathrm{s}, t\}$, which is defined from a mathematical perspective [115].

1. Identity:

$$\Theta(\mathrm{s}^*, \mathrm{s}') = 0, \text{if and only if } \mathrm{s}^* = \mathrm{s}' \tag{5.7}$$

2. Symmetry:

$$\Theta(\mathrm{s}^*, \mathrm{s}') = \Theta(\mathrm{s}', \mathrm{s}^*) \tag{5.8}$$

3. Triangle inequality:

$$\Theta(\mathrm{s}^*, \mathrm{s}') \leq \Theta(\mathrm{s}^*, \mathrm{s}^-) + \Theta(\mathrm{s}^-, \mathrm{s}') \tag{5.9}$$

5.2.3 CONCEPTUAL SOLUTION

In this subsection, let $\mathrm{s}^* \in \{\mathrm{s}, t\}$ denote the ground truth and $\mathrm{s}' \in \{\mathrm{s}, t\}$ represent the actual state during the process of the system evolution. We define the comprehensive performance metric as $\Theta(\mathbf{s}^*,\mathbf{s}')$. It can assess the overall performance quantitatively, which suggests that this metric can integrate multiple performance indexes into a single unified one, providing a holistic assessment.

This subsection defines a pairing $< s_j, \tilde{s}_j >$ to represent relevant spatio-temporal signals in $\{\mathbf{s}, t\}$, and $\tilde{s}_j$ is calculated as a specific function of s_j and t. This pairing establishes a direct relationship between the system states and their corresponding time-derived values. In this chapter, the ground truth set $\mathbf{s}^*$ represents the expected states that can be derived from the predesigned system objectives, which is expressed as follows.

$$\mathbf{s}^* = \{< s_1^*, \tilde{s}_1^* >, < s_2^*, \tilde{s}_2^* >, \ldots, < s_q^*, \tilde{s}_q^* >\} \tag{5.10}$$

Moreover, a similar formulation of the actual dynamic states can be given by

$$\mathbf{s}' = \{< s_1', \tilde{s}_1' >, < s_2', \tilde{s}_2' >, \ldots, < s_n', \tilde{s}_n' >\} \tag{5.11}$$

where the cardinalities of sets $\mathbf{s}^*$ and $\mathbf{s}'$ are q and n, respectively ($q \geq n$). $\mathbf{s}^*$ consists of a broader of dynamic states compared to $\mathbf{s}'$ such that sufficient reference points can be obtained for comparison.

Based on the above definition and discussion, we can introduce the definition of the comprehensive metric in a rigorous and interpretable way [28].

Definition 5.1 *Given the ground truth* $\mathbf{s}^*$ *as the baseline, we define the quantitative metric of the system's dynamic states* $\mathbf{s}'$ *by*

$$\Theta(\mathbf{s}') = \left\{ \left[\sum_{i=1}^{n} \left(\theta_{\delta_i} \left(< s'_i, \tilde{s}_i' >, < s_i^*, \tilde{s}_i^* > \right) \right)^{\nu'} + \sum_{j=1}^{q-n} \left(\delta_j^{\xi} \cdot \xi \right)^{\nu'} \right] \frac{1}{q} \right\}^{1/\nu'} \tag{5.12}$$

$$\theta_{\delta_i} \left(< s'_i, \tilde{s}_i' >, < s_i^*, \tilde{s}_i^* > \right) = \left[\delta_{i1} \cdot d(s'_i, s_i^*)^{\nu'} + \delta_{i2} \cdot d(\tilde{s}_i', \tilde{s}_i^*)^{\nu'} \right]^{1/\nu'} \tag{5.13}$$

where $\theta_{\delta_i} (< s'_i, \tilde{s}_i' >, < s_i^*, \tilde{s}_i^* >)$ *represents the base distance between the dynamic state set and the corresponding ground truth. Additionally,* ξ *denotes the compensated distance, which can compensate for the disparities in the state comparisons. The* ν'*-norm is adopted as the distance function* $d(*, *)$ *in this chapter.* $\nu'(\nu' \geq 1)$ *is the metric order parameter, which determines the sensitivity.* δ_i *and* δ_j^{ξ} *control the weightings of relevant factors, where* $\delta_i = [\delta_{i1}, \delta_{i2}]$*, and similarly,* δ_j^{ξ} *has* $q - n$ *elements.*

As stated previously, the cardinalities p and n satisfy the condition $p \geq n$. For this reason, the compensated distance ξ is interpreted as a measure of the inaccessible dynamic states which are not directly observable or cannot be effectively matched to the corresponding ground truth states. The order parameter ν' plays a critical role in determining the relative weighting of "outlier" dynamic states, which deviate greatly from their corresponding ground truth values. By adjusting ν', the metric can emphasize or de-emphasize the impact of these outliers, providing flexibility in the performance evaluation. Furthermore, the parameters δ_i and δ_j^{ξ} are designed to control the contributions of different states to the overall metric. Concretely, δ_i influences the weightings of the evaluated states, while δ_j^{ξ} adjusts the contributions of the compensated states. All these parameters ensure that the metric can accurately reflect the importance of various dynamic states, tailoring the evaluation to the corresponding characteristics of the system.

5.2.4 METRIC DESIGN FOR THE PERFORMANCE EVALUATION OF CHASSIS DOMAIN DYNAMIC

This chapter has presented the performance-relevant dynamic states as well as their derivatives, which is demonstrated in Fig. 5.4. This subsection will utilize these spatio-temporal signals to design the performance metric for chassis domain dynamics.

The associated variables and their corresponding derivatives are paired, as expressed in Eq. (5.14). These pairings provide the foundation for quantifying chassis performance in an organized and consistent manner. In addition, the ground truth states for chassis dynamic signals can be intuitively and reasonably set to zero. This assumption implies that a relatively stationary vehicle operates with minimal deviations in its dynamic states, which represents the optimal scenario for chassis dynamic performance. Such a baseline accords with the desired notion of handling stability and dynamic responses, facilitating easy interpretation and comparison.

$$\begin{aligned}\mathbf{s}_v =& \{< a_x, \dot{a}_x >, < \lambda, \dot{\lambda} >, < \dot{\theta}_v, \ddot{\theta}_v >, < a_y, \dot{a}_y >,\\ & < \beta, \dot{\beta} >, < \varphi, \dot{\varphi} >, < \mathrm{LTR}, d\mathrm{LTR}/dt >, < \dot{\psi}_r, \ddot{\psi}_r >\}\end{aligned} \tag{5.14}$$

To ensure consistency and comparability of various dynamic states, all involved signals are normalized according to the analysis of their boundaries. The details of their normalization factors are discussed in Section 5.2. Notably, this chapter uniformly defines the normalization factors for all signal derivatives as one. In the following, the original notations of all selected signals are used to correspond to their normalized forms. This normalization ensures that the designed metric remains robust and interpretable, regardless of the scale or unit of the original signals. Then, we can yield the formulation of the comprehensive performance metric based on Eqs. (5.12) and (5.13). It is worth mentioning that hyperparameters play an important role in refining the metric to reflect the relative importance of various factors influencing chassis dynamic performance. To further enhance this metric, we have to investigate the selection and setting of key hyperparameters, such as the weightings associated with involved signals and derivatives.

The compensated distance ξ represents the metric factor that cannot be directly assigned. This chapter determines its value through the correlation analysis of vehicle dynamic states, which is demonstrated in Fig. 5.3. The relationships between various dynamic states including both measurable and inaccessible states can be used to deal with the informed estimations of inaccessible states. Weightings derived from the correlation analysis imply the degree of association between measurable and non-measurable states. This chapter assigns related available states to the inaccessible dynamic states based on these weightings, which are meaningful and grounded in empirical data. Thus, the proposed metric can deal with scenarios where some dynamic states can not be directly measured by naturally extending to inaccessible states, making it suitable for application in mass-production vehicles.

The order parameter ν' influences how weights are assigned to large values among chassis dynamic signals, which plays a crucial role in the proposed performance evaluation method. A larger ν' assigns greater weight to larger signal values, and it will make the metric more sensitive and less tolerant to significant deviations. Concretely, this chapter highlights two typical settings of ν':

1. $\nu' = 1$: In this case, the proposed metric measures a first-order status of the involved states, which corresponds to the simple sum of all performance-related components. This setup provides a straightforward interpretation, and it makes the metric easily understandable and practical for scenarios requiring direct and transparent performance evaluations.
2. $\nu' = 2$: This option is more practical since it produces smoother data while increasing sensitivity to "outliers" or states with larger deviations. This makes it effective in detecting and representing atypical dynamic behaviors.

Furthermore, the weightings of individual states are determined based on their classification directions, and thus the developed metric could reflect the contributions of different state types. By incorporating this weighted method, the proposed metric can be scalable to a wide range of vehicle types and application scenarios. The proposed methodology can maintain versatility and robustness via the fine-tuning of key hyperparameters, and we argue that it is a good tool for evaluating chassis domain dynamics in various scenarios.

5.2.5 RESULTS AND ANALYSIS

5.2.5.1 Hyperparameters Analysis

The metric distribution with a different order parameter ν' is demonstrated in Fig. 5.7, which validates the statement highlighted in the previous subsection. In the case of $\nu' = 1$, the metric values are dispersed between zero and

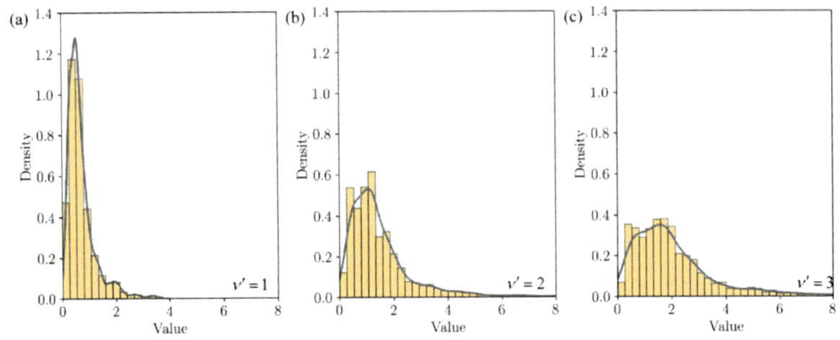

Figure 5.7 Metric distributions with different order parameters.

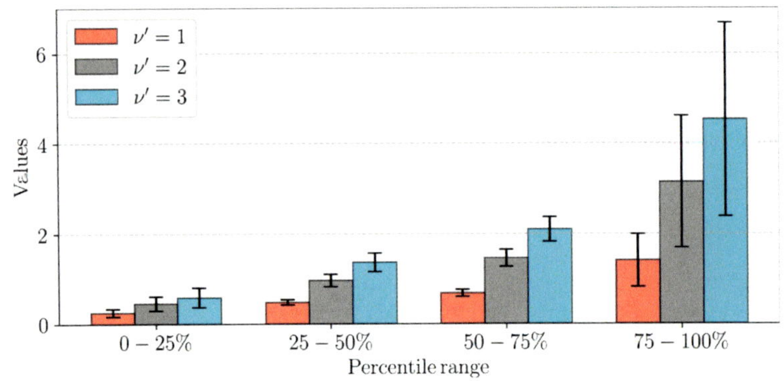

Figure 5.8 Metric distribution statistics of different ν' values.

one, which suggests the straight aggregation of various performance-related components. However, as the order parameter increases, the peak value of the proposed metric also rises, and the metric distribution becomes smoother. Specifically, the peak values of the metric are 3.7024 ($\nu' = 1$), 9.1717 ($\nu' = 2$), and 13.5512 ($\nu' = 3$), while the corresponding mean values are 0.7032, 1.5025, and 2.1431, respectively. This trend demonstrates that larger order parameters could improve the metric's sensitivity to signal variations.

Fig. 5.8 presents the mean values and their corresponding standard deviations (SD) of different intervals, which also implies that a larger ν' leads to greater sensitivity in the performance metric and facilitates to capture finer variations in dynamic states. The proposed assessment method for chassis domain performance has the attractive potential to analyze driving styles. It is observed that a smaller order parameter results in a smaller metric value for the same dynamic status. Thus, we can combine different driving styles with corresponding order parameters, for example, a more tolerant metric (a smaller ν') is better suited for evaluating a more aggressive driving style. Thus, the proposed metric can not only provide a robust and sensitive

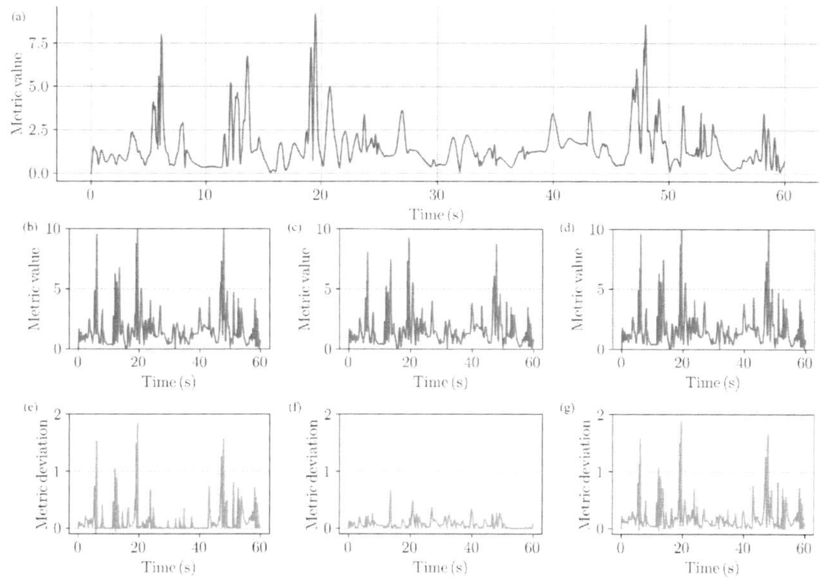

Figure 5.9 Metric values and deviations in typical scenarios of signal inaccessibility.

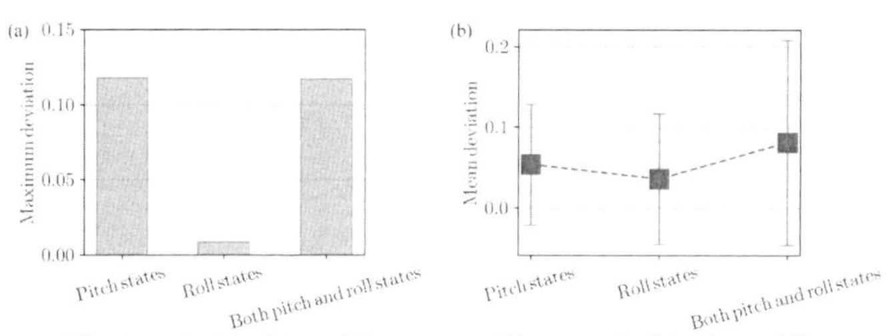

Figure 5.10 Metric deviation statistics in typical scenarios of signal inaccessibility.

measure of chassis domain dynamics but also enable a deeper understanding of driving behaviors.

Figs. 5.9 and 5.10 illustrate the metric deviations and statistics in different scenarios of signal inaccessibility, which suggests the effectiveness of the proposed metric in evaluating chassis domain dynamics. It can be found that the proposed quantitative performance metric is robust against signal losses so that it has practical applicability and reliability regardless of some inaccessible signals in mass-production cars. As presented in Fig. 5.9, the metric values are consistent in different signal-loss scenarios, and this consistency highlights the resilience of the proposed metric to missing signals.

It can be observed from Fig. 5.10 that:

1. Pitch state is inaccessible: The mean value of metric deviations is -0.0537 ± 0.0750, and the maximum deviation achieves 0.1179 in this case.
2. Both pitch and roll states are inaccessible: For this worst-case scenario, the mean value of metric deviations increases slightly to -0.0812 ± 0.1271. The maximum deviation is 0.1172.

All these results suggest that the proposed metric can effectively compensate for inaccessible dynamic states and is robust against challenging conditions with losses of some dynamic states. Such robustness enhances the practicality of the metric for evaluating vehicle dynamics in various environments and applications.

5.2.5.2 Large Acceleration Scenario

Fig. 5.11 and Fig. 5.12 demonstrate the experimental results of the large acceleration scenario. The vehicle operates with large acceleration, using a large throttle opening. Fig. 5.11 shows the longitudinal acceleration and speed of the test car, and it shows that the longitudinal acceleration reaches nearly 10 m/s^2 within the first second, suggesting that the vehicle reaches its road grip limit. This situation might push the test vehicle to lose dynamic stability, and the proposed metric can quantitatively indicate this dangerous condition, as shown in Fig. 5.12. It can be observed that the metric value fluctuates with the change of vehicle states, which can represent chassis domain performance in real-time. We can see that the metric value reaches its peak at about 0.5 s,

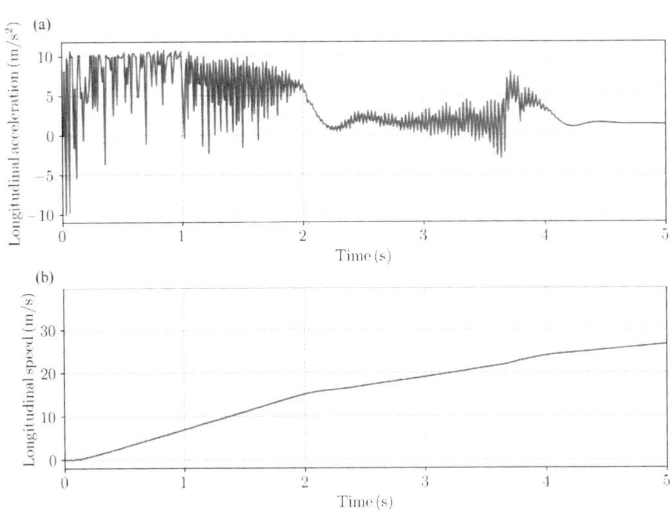

Figure 5.11 Dynamic states in the large acceleration scenario, (a) longitudinal acceleration, (b) longitudinal speed.

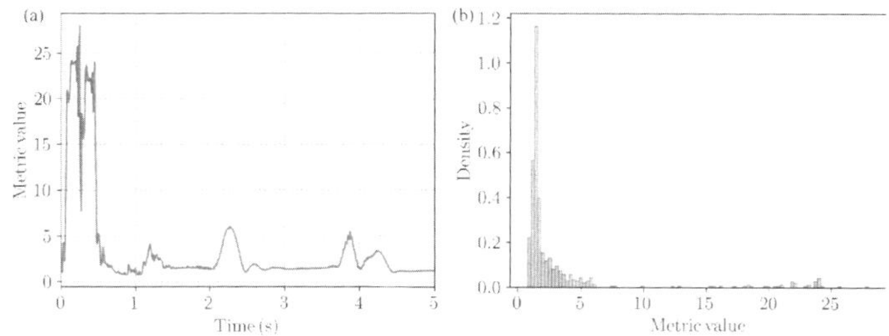

Figure 5.12 Metric values in the large acceleration scenario, (a) metric curve, (b) metric distribution.

and its value is about 20, which indicates the test car's dangerous status. The metric value then decreases to about two with the decrease of vehicle longitudinal acceleration, which suggests that the test car returns to a safe and suitable driving status.

5.2.5.3 Double Lane Change Scenario

Fig. 5.13 and Fig. 5.14 illustrate the experimental results of the double lane change scenario. The test car performs the double lane change maneuver at a speed of about 80 km/h, and vehicle dynamic states are presented in Fig. 5.13. This test scenario is designed to evaluate the proposed metric in a situation where the vehicle operates with lateral maneuvers. Fig. 5.14 shows the metric values in the double lane change scenario, and it can be found that the metric

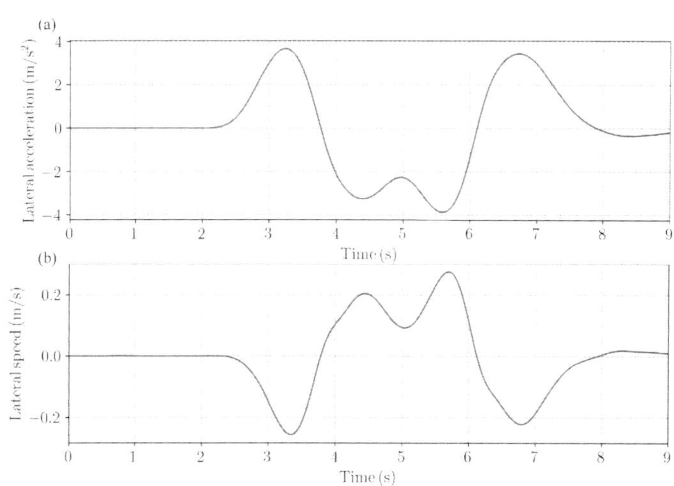

Figure 5.13 Vehicle dynamics states in the double lane change scenario, (a) longitudinal acceleration, (b) longitudinal speed.

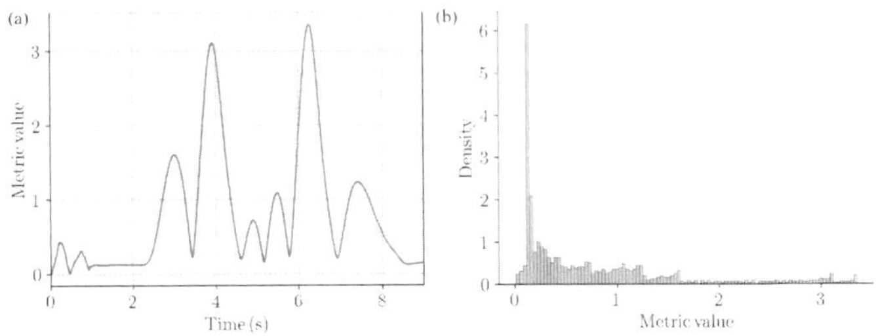

Figure 5.14 Metric values in the double lane change scenario, (a) metric curve, (b) metric distribution.

values fluctuate with the change of dynamic states. It can be observed that the metric value reaches its peak at about 4s and 6.5s, respectively, and the peak values are about three, which indicates that the vehicle approaches the vehicle lateral dynamic limits during the double lane change maneuver. While the metric value is less than one most of the time.

The metric variations closely follow the trends in vehicle dynamic signals including vehicle lateral speed and lateral acceleration. This phenomenon can be attributed to the variations of steering wheel angles. The results in this scenario demonstrate the effectiveness of the proposed metric, particularly in assessing the performance of vehicle lateral and roll dynamics.

5.2.5.4 Road Course Scenario

In this test scenario, the test car runs on a typical highway road, and vehicle dynamic states are presented in Figs. 5.15 and 5.16. Besides, the metric values are shown in Fig. 5.17, and it can be found that the metric values fluctuate with the change of dynamic states. It is worth mentioning that the road course scenario is designed to evaluate the proposed metric in a situation where the car operates with various lateral and longitudinal maneuvers. The metric value keeps less than two most of the time, which suggests that the test car

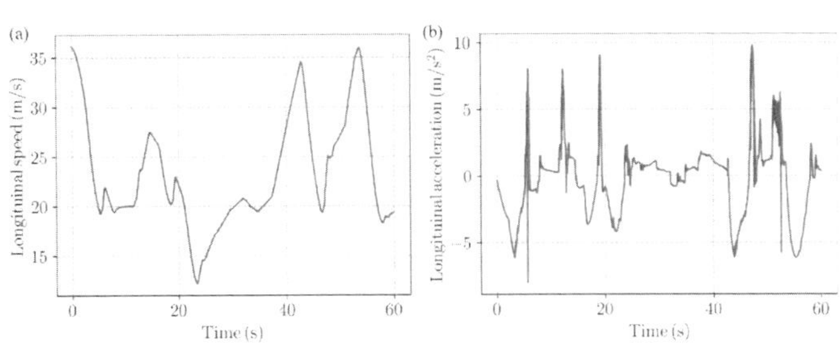

Figure 5.15 Dynamic states in the road course scenario, (a) longitudinal speed, (b) longitudinal acceleration.

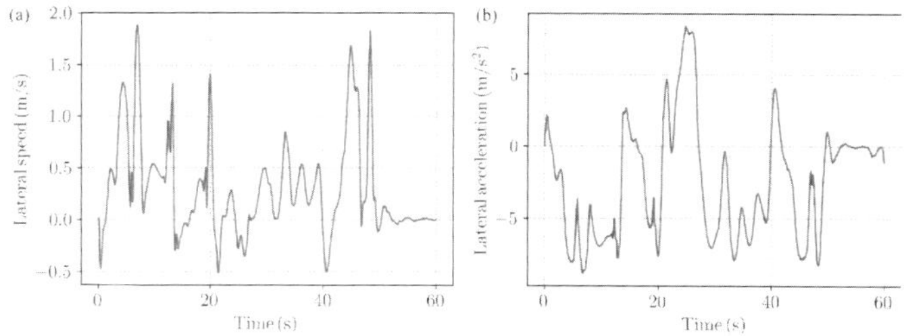

Figure 5.16 Dynamic states in the road course scenario, (a) lateral speed, (b) lateral acceleration.

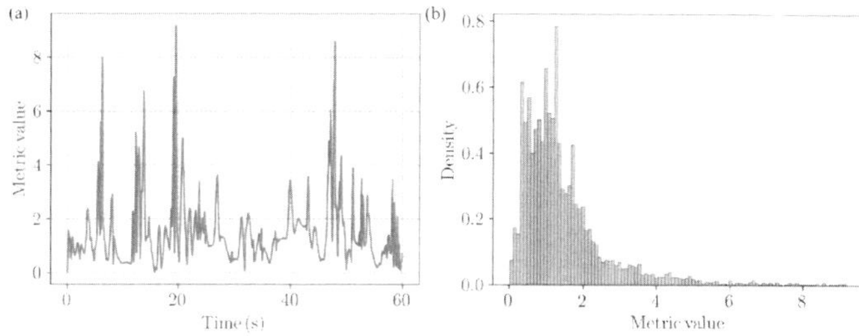

Figure 5.17 Metric curve in the road course scenario, (a) metric curve, (b) metric distribution.

remains a favorable dynamic status in the road course scenario. Furthermore, the exhibited fluctuations of the metric curve reflect the real-time performance of chassis domain dynamics. Four peaks are observed in the metric value curve. Specifically, the metric achieves larger values at about 7 s, 13 s, 19.5 s, and 47 s, respectively, and the peak values are all larger than five, which indicates that the test car approaches vehicle dynamic limits. Interestingly, these peaks in the metric curve align approximately in time with similar peaks in the longitudinal acceleration curve.

It can be observed from comparisons of Figs. 5.15, 5.16, and 5.17 that the sharp changes, especially those that lead to abrupt shifts in vehicle speed, are the main causes of discomfort for the occupants, as indicated by larger metric values. This verifies the effectiveness of the proposed performance metric in assessing vehicle ride comfort, especially in situations where rapid acceleration or deceleration induces discomfort for the occupants. In conclusion, the proposed metric can effectively capture features of chassis domain dynamics and provide insights into both riding comfort and driving safety in real-world driving conditions.

5.2.5.5 Metric Statistics for Various Scenarios

This chapter also conducts and analyzes a series of typical test scenarios to verify the effectiveness and robustness of the proposed assessment approach.

Table 5.2
Statistics of performance metric in various test scenarios.

Vehicle speed	Road friction coefficient		
	$\mu = 0.6$	$\mu = 0.8$	$\mu = 1.0$
60 km/h	2.1284 ± 0.3184	2.1003 ± 0.2730	2.0844 ± 0.2457
80 km/h	2.9946 ± 0.5115	2.8740 ± 0.4119	2.8129 ± 0.3629
100 km/h	4.3594 ± 0.9481	3.984 ± 0.7014	3.7922 ± 0.5902

The double lane change test scenario is performed with different vehicle speeds and road adhesion frictions to examine the metric's adaptability and sensitivity to various driving conditions. When a car operates at higher speeds or on a lower friction road, it is more likely to lose dynamic stability and occur dangerous situations. Table 5.2 presents the mean values and standard deviations of the top 10% of metric values in different test conditions. It can be observed that the metric values increase with the increase of vehicle speed and decrease of road friction coefficient, which suggests that the proposed metric can effectively assess the impact of vehicle speeds and road friction coefficients in various driving conditions. Specifically, when the test car runs at a speed of 60 km/h, the metric values achieve the peak values of 2.1284, 2.1003, and 2.0844 with road friction coefficients 0.6, 0.8, and 1.0, respectively. A similar trend can be observed at speeds of 80 and 100 km/h. Moreover, the descent gradient of metric value with respect to road friction coefficient increases with the increase of vehicle speed, which suggests that the proposed metric is more sensitive to road friction at higher speeds. The reason is that when a car runs at relatively lower speeds, the vehicle has smaller road adhesion utilization, allowing it to maintain adequate dynamic performance across various road conditions. As speed increases, the vehicle's dependency on road adhesion grows, which highlights the significance of road friction in guaranteeing stable and safe driving performance.

5.2.6 SUMMARY

This section proposes a novel quantitative performance metric for the evaluation of chassis domain dynamics. Both dynamic stability and riding comfort are considered in the proposed metric, which can provide a comprehensive and objective assessment. This chapter formulates the general problem of comprehensive performance evaluation of vehicle multiple performance and defines the essential properties of the performance metric. The metric is designed to quantify the "difference" between two sets of dynamic states, which can be both mathematically consistent and physically interpretable. A variety of experimental results validate the effectiveness of the proposed metric, including the road course test, double lane change test, and large acceleration test. This metric can be used as a reliable tool for quantifying the levels of vehicle dynamic stability and riding comfort. Specifically, the metric can be directly applied as a control objective or a constraint to optimize the trajectory planning module and control algorithms.

5.3 PROBLEM STATEMENT OF PATH PLANNING

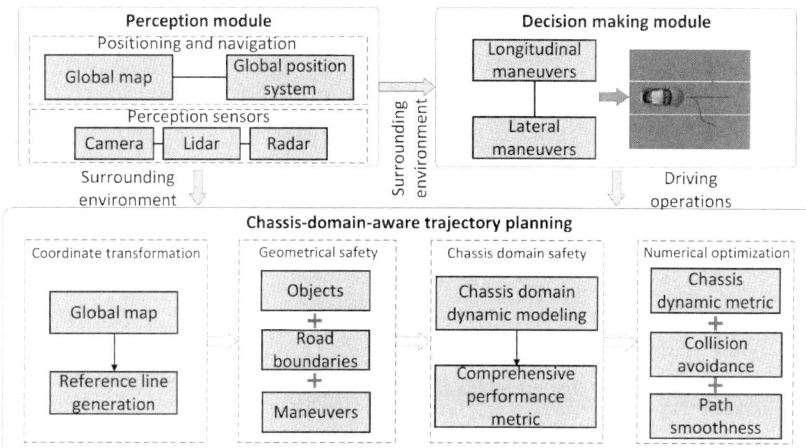

Figure 5.18 The diagram of the proposed trajectory planning method.

The overall framework is illustrated in Fig. 5.18, which involves three main modules: the perception module, the decision-making module, and the trajectory planning module. The perception module detects environmental information, including the surrounding obstacles, and road lanes. The positioning and navigation system detects vehicle positions and determines the desired routes. The decision-making module generates driving commands based on the detected environmental information and the desired routes. Then, the trajectory planning module optimizes desired local trajectories based on various aspects of performance.

As demonstrated in Fig. 5.18, the global map is used to generate the reference line via the coordinate transformation. The proposed trajectory planning method formulates geometric safety by considering road geometric constraints and surrounding vehicles. The dynamical stability is ensured by the dynamical stability checker, which guarantees the vehicle's handling stability. The proposed performance metric for chassis domain dynamics is employed to optimize the trajectory profiles. Thus, we can perform the trajectory planning by solving the numerical optimization problem concerning the chassis domain dynamics, feasible safety, and geometrical smoothness. This

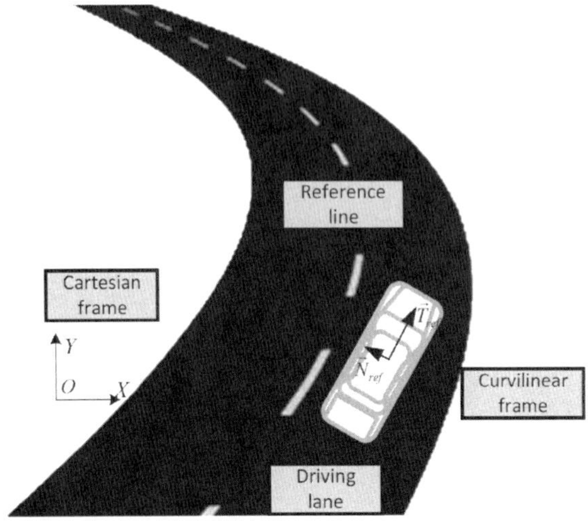

Figure 5.19 Demonstration of the Cartesian coordinate and curvilinear coordinate.

optimization problem also involves two safety constraints: the geometric safety checker and the dynamic stability checker.

The designed trajectory planning approach utilizes a moving curvilinear coordinate to perform the trajectory optimization. This coordinate is constructed based on the reference line extracted from the global map, which is shown in Fig. 5.19. Specifically, the tangential vector $\vec{T}_{ref}$ and normal vector $\vec{N}_{ref}$ at a certain point along the reference line are used to construct the coordinate. In this chapter, we employ the cubic spline interpolation, which can generate smooth curves through a number of waypoints, to represent the reference line. Basically, the reference line is represented by a set of waypoints in the Cartesian coordinate, and the cubic spline interpolation is used to fit the reference line in the curvilinear coordinate. Assuming that n_w waypoints with coordinates $(x_{w,i}, y_{w,i})$ in the Cartesian coordinate are obtained, this chapter formulates piece-wise cubic polynomials $[\Upsilon_x(s_{ref}), \Upsilon_y(s_{ref})]^T$, which are defined over the arc length s_{ref} of the reference line. These polynomials are used to depict the reference line in the curvilinear frame. The expression is presented as follows.

$$\begin{aligned} \begin{bmatrix} \Upsilon_x(s_{ref}) \\ \Upsilon_y(s_{ref}) \end{bmatrix} &= \begin{bmatrix} \gamma_{x,i}(s_{ref}) \\ \gamma_{y,i}(s_{ref}) \end{bmatrix}, \\ \text{with} \quad & s_{ref,i} \leq s_{ref} < s_{ref,i+1} \end{aligned} \tag{5.15}$$

$$\begin{bmatrix} \gamma_{x,i}(s_{ref}) \\ \gamma_{y,i}(s_{ref}) \end{bmatrix} = \begin{bmatrix} \varpi_{x3} & \varpi_{y3} \\ \varpi_{x2} & \varpi_{y2} \\ \varpi_{x1} & \varpi_{y1} \\ \varpi_{x0} & \varpi_{y0} \end{bmatrix}^T \begin{bmatrix} (s_{ref} - s_{ref,i})^3 \\ (s_{ref} - s_{ref,i})^2 \\ s_{ref} - s_{ref,i} \\ 1 \end{bmatrix} \tag{5.16}$$

where $i = 1, 2, ..., n_w - 1$. This section utilizes the mathematical spline principle to calibrate the coefficients ϖ_{xi} and ϖ_{yi} such that the fitted line can pass through each data point without any irregularities or breaks in continuity [94]. Thus, we can obtain reference bases to perform the trajectory planning optimization.

The target path point P_d is defined in terms of the parameters (s_d, l_d) in the curvilinear coordinate system. In particular, l_d is the lateral offset from the root point P_r in the direction of $\vec{N}_{ref}$, which leads to the desired path point P_d. The root point P_r is the projection point of P_d in the curvilinear frame. s_d is the arc length from the origin point to the root point P_r.

This coordinate representation provides a natural and intuitive way to link the desired path point to vehicle longitudinal and lateral motions. Thus, the initial states for path planning can be determined by mapping the vehicle's current states. This chapter assumes that all the required inputs from the decision-making module are available. Based on the driving commands required by the decision-making module, we can yield the terminal states of the path planning problem. Thus, the optimization procedure ensures that the terminal state aligns seamlessly with the objectives required by these high-level commands.

5.4 CHASSIS-DOMAIN-AWARE TRAJECTORY PLANNING

Quartic and quintic polynomials in the curvilinear coordinate are utilized to represent the local paths for the longitudinal direction and lateral direction. First, this section defines the initial state as $P_{in} \Rightarrow [s_{in}, \dot{s}_{in}, \ddot{s}_{in}, l_{in}, \dot{l}_{in}, \ddot{l}_{in}]$ at t_{in} and defines the terminal state as $P_{te} \Rightarrow [s_{te}, \dot{s}_{te}, \ddot{s}_{te}, l_{te}, \dot{l}_{te}, \ddot{l}_{te}]$ at t_{te} with the planning time horizon t_f. Then, we utilize $\Omega(t)$ to represent the planned trajectory profiles in the curvilinear coordinate. The resulting trajectory equations are required to integrate the boundary conditions imposed by P_{in} and P_{te}, which guarantees the kinematic and dynamic continuity. The formulations of $\Omega(t)$ are given as follows.

$$\Omega(t) : \begin{cases} s_d(t) = \rho_{s4}t^4 + \rho_{s3}t^3 + \rho_{s2}t^2 + \rho_{s1}t + \rho_{s0} \\ l_d(t) = \rho_{l5}t^5 + \rho_{l4}t^4 + \rho_{l3}t^3 + \rho_{l2}t^2 + \rho_{l1}t + \rho_{l0} \end{cases} \tag{5.17}$$

where ρ_{si} and ρ_{li} are polynomial coefficients.

The hybrid state space is defined as $\boldsymbol{x} = [s_d, v_x, a_x, j_x, l_d, v_y, a_y, j_y, \beta, \varphi, t]^T$, which combines crucial vehicle dynamic states and the trajectory formulation in Eq. (5.17). Moreover, j_x and j_y denote the derivative of a_x and a_y, respectively. This state space can provide a holistic representation of vehicle motion and dynamic performance, including both kinematic and dynamic aspects of the vehicle's behavior. This comprehensive representation is essential for trajectory planning optimization that balances dynamic stability, riding comfort, and complicated surrounding environments.

To yield the hybrid ordinary differential equations for the local path planning problem, the following reasonable assumptions are introduced to simplify the complex vehicle dynamics:

1. **Exact trajectory tracking:** The vehicle is assumed to track the desired trajectory without any deviations. This simplifies the kinematic relationship between the vehicle and the reference path. This assumption implies that the vehicle's local coordinate aligns seamlessly with the curvilinear coordinate, which is necessary and reasonable for local trajectory planning optimization.
2. **Uniform longitudinal tire forces:** The longitudinal tire forces on the left and right tires of the same axle are assumed to be equal to each other. Besides, these forces are assumed to be proportional to the corresponding vertical tire forces. This assumption reduces tire force components and simplifies the mathematical formulations of the path planning problem.

Based on the aforementioned assumptions, the evolution formulation of the dynamic system can be expressed as follows, wherein the trajectory states and vehicle dynamics are integrated into a unified dynamic system.

$$\dot{\boldsymbol{x}} = f(\boldsymbol{x}, \boldsymbol{u}_\rho, \boldsymbol{u}_v) \tag{5.18}$$

where $\boldsymbol{u}_\rho = [\rho_{l5}, \rho_{l4}, \rho_{s4}]^T$ and $\boldsymbol{u}_v = [F_{xf}, F_{xr}, \delta]^T$. In particular, the remaining polynomial coefficients, which fall outside the control input $\boldsymbol{u}_\rho$, are calculated based on the initial states.

The equations describing the evolution of the unified dynamic system are summarized as follows.

$$\dot{x}_i = x_{i+1}, \quad i \in [1, 2, 3, 5, 6, 7], \tag{5.19a}$$

$$\dot{x}_4 = 24 \cdot u_{\rho,3}, \tag{5.19b}$$

$$\dot{x}_8 = 120 \cdot u_{\rho,1} \cdot x_{11} + 24 \cdot u_{\rho,2}, \tag{5.19c}$$

$$\dot{x}_9 = x_7 / x_2, \tag{5.19d}$$

$$\begin{aligned} \dot{x}_{10} = {} & [l_r \cdot k_r - l_f \cdot k_f \cdot \cos u_{v,3}] x_6 / [I_z \cdot x_2] \\ & - [l_r^2 \cdot k_r + l_f^2 \cdot k_f \cdot \cos u_{v,3}] x_{10} / [I_z \cdot x_2] \\ & + [l_f \cdot k_f \cdot u_{v,3} \cdot \cos u_{v,3} + l_f \cdot u_{v,1} \cdot \sin u_{v,3}] / I_z, \end{aligned} \tag{5.19e}$$

$$\dot{x}_{11} = 1, \tag{5.19f}$$

where k_f and k_r denote the front and rear axles' cornering stiffnesses, respectively.

The fundamental objective of the trajectory planning problem lies in the optimization of the polynomial coefficients ρ_{si} and ρ_{li}, according to the designed trajectory profiles in Eq. (5.17). In addition, some of these coefficients can be calculated based on the initial states. This dependency can be observed in Eq. (5.19) which involves $\boldsymbol{u}_\rho$. The optimization procedure utilizes this relationship so that the planned trajectory seamlessly integrates with the vehicle's current state while satisfying dynamic constraints.

However, the integration of vehicle dynamics and trajectory planning introduces extra variables, i.e., $\boldsymbol{u}_v$, which represent control inputs related to vehicle dynamics. In this section, these variables can be resolved by utilizing the vehicle's physical constraints, which are expressed by the equation $h_v(\boldsymbol{x}, \boldsymbol{u}_v) = 0$. According to the vehicle dynamic modeling in Section 2.4, we can yield the detailed formulations of $h_v(\boldsymbol{x}, \boldsymbol{u}_v) = 0$, which are expressed as follows.

$$u_{v,1}[g \cdot l_f + x_3 \cdot h_g] - u_{v,2}[g \cdot l_f - x_3 \cdot h_g] = 0, \tag{5.20a}$$

$$\begin{aligned} &- k_f \cdot \sin u_{v,3} \left[u_{v,3} - \frac{x_6 + l_f \cdot x_{10}}{x_2} \right] + u_{v,2} \\ &+ u_{v,1} \cdot \cos u_{v,3} - m[x_3 - x_6 \cdot x_{10}] = 0, \end{aligned} \tag{5.20b}$$

$$\begin{aligned} &k_f \cdot \cos u_{v,3} \left[u_{v,3} - \frac{x_6 + l_f \cdot x_{10}}{x_2} \right] + u_{v,1} \cdot \sin u_{v,3} \\ &- k_r \frac{x_6 - l_r \cdot x_{10}}{x_2} - m[x_7 + x_2 \cdot x_{10}] = 0, \end{aligned} \tag{5.20c}$$

The local path planning task can thus be constructed as a nonlinearly constrained optimization problem. The cost function involves the following components:

1. **Reference line cost:** This component penalizes deviations from the desired reference line to keep adherence to the desired path.
2. **Vehicle dynamics cost:** This term is designed to improve vehicle dynamic performance.
3. **Path smoothness cost:** This component is designed to ensure that the planned trajectory is smooth and continuous.
4. **Collision risk cost:** The collision risk term minimizes the likelihood of collisions with surrounding obstacles.

A set of constraints of the path planning problem is also introduced as follows.

1. **Dynamic system evolution:** This constraint term ensures state transitions governed by the hybrid ordinary differential equations.
2. **Vehicle physical equations:** This constraint enforces the physical evolution of the vehicle dynamic system.
3. **Road geometry constraints:** This constraint ensures that the planned trajectory satisfies the road geometry and remains within the road boundaries.
4. **Collision avoidance constraints:** This component guarantees safe distances from surrounding cars and obstacles.

5. **Vehicle handling limits:** This component enforces chassis dynamic stability constraints to maintain vehicle maneuverability.
6. **Initial and terminal state constraints:** This component ensures that the path satisfies specific goals and conditions at the beginning and end of the planning horizon.

5.4.1 COST FUNCTION

5.4.1.1 Reference Line Cost

The reference line is predefined based on the control command from the decision-making module. The local path planning task is required to ensure adherence to the desired reference line as much as possible. Therefore, this section designs a quadratic cost function $\Xi_r(\boldsymbol{x})$ to penalize deviations from the reference line, which is expressed as follows.

$$\Xi_r(\boldsymbol{x}) = \|\boldsymbol{x} - \boldsymbol{x}_{ref}\|^2_{Q_r} \tag{5.21}$$

$$\boldsymbol{x}_{ref} = [0, v_{x,ref}, 0, 0, l_{ref}, \underbrace{0, ..., 0}_{6}] \tag{5.22}$$

where Q_r is the weighting matrix of this cost component.

5.4.1.2 Vehicle Dynamics Cost

Similar to the reference line cost, we can design the quadratic function $\Xi_v(\boldsymbol{x})$ to represent vehicle dynamics cost based on key vehicle dynamic states.

$$\Xi_v(\boldsymbol{x}) = \|[x_3, x_7, x_9, x_{10}]\|^2_{Q_v} \tag{5.23}$$

where Q_v represents the weighting matrix.

In addition, we can adopt the comprehensive performance metric for chassis domain dynamics which has been introduced in the former section, as the cost component for vehicle dynamics. This metric provides a holistic evaluation of chassis domain performance.

5.4.1.3 Path Smoothness Cost

The path smoothness cost $\Xi_p(\boldsymbol{x})$ is introduced to penalize lateral and longitudinal jerks. It can be expressed as a quadratic function as follows.

$$\Xi_p(\boldsymbol{x}) = \|[x_4, x_8]\|^2_{Q_p} \tag{5.24}$$

where Q_p is the weighting matrix that determines the importance of this cost component.

5.4.1.4 Collision Risk Cost

The collision risk cost $J_c(\boldsymbol{x})$ can quantify the proximity of the planned trajectory to surroundings via the Euclidean distance between the ego car and its surroundings, so that it can be used to make the ego car keep a safe distance from potential collisions. It is defined as

$$J_c(\boldsymbol{x}) = \min\{\|[x_1 - s_z^{\text{obs}}, x_5 - l_z^{\text{obs}}]\|^2, \forall(s_z^{\text{obs}}, l_z^{\text{obs}}) \in \boldsymbol{C}_{\text{obs}}\} \tag{5.25}$$

where $(s_z^{\text{obs}}, l_z^{\text{obs}})$ corresponds to the longitudinal and lateral positions of an obstacle vehicle and $\boldsymbol{C}_{\text{obs}}$ is the set of all obstacles.

5.4.2 MULTIPLE CONSTRAINTS

First, it is essential to impose the constraints of dynamic system evolution (Shown in Eq. (5.19)), which are expressed as

$$\dot{\boldsymbol{x}} = f(\boldsymbol{x}, \boldsymbol{u}_\rho, \boldsymbol{u}_v) \tag{5.26}$$

Given that the proposed trajectory optimization problem involves vehicle physical dynamics, the equations of the vehicle physical model (Presented in Eq. (5.20)) are involved as a constraint, which is expressed as

$$h_v(\boldsymbol{x}, \boldsymbol{u}_v) = 0 \tag{5.27}$$

To guarantee collision avoidance when performing local path planning, we introduce an elliptic function $\psi_c(\boldsymbol{x}) > 1$. This function establishes a safe boundary around obstacles to keep the planned trajectories outside these boundaries. The constraint is expressed as

$$\begin{aligned}\psi_c(\boldsymbol{x}) &= \left[x_1 - s_z^{\text{obs}}, x_5 - l_z^{\text{obs}}\right] Q_c \left[x_1 - s_z^{\text{obs}}, x_5 - l_z^{\text{obs}}\right]^T \\ &> 1, \forall(s_z^{\text{obs}}, l_z^{\text{obs}}) \in \boldsymbol{C}_{\text{obs}}\end{aligned} \tag{5.28}$$

where Q_c represents the elliptic weighting matrix, and it defines the shape of the elliptic boundary, which accounts for the size of the vehicle and the safety distance.

Then, we introduce one constraint of vehicle handling limits, $\psi_t(\boldsymbol{x}) < 0$, which limits the vehicle's handling within the physical limits of the tire friction. This constraint can be expressed as

$$[x_3 - x_6 \cdot x_{10}]^2 + [x_7 + x_2 \cdot x_{10}]^2 - (\mu \cdot g)^2 < 0 \tag{5.29}$$

On the other hand, we analyze the equilibrium of tire forces when the front or rear tires are operating at their saturation limits, as introduced in the previous literature [28]. We can yield the boundaries of vehicle lateral responses based on this analysis, which can be utilized as a constraint to ensure vehicle handling performance. This constraint, namely $\psi_l(\boldsymbol{x}) < 0$, is expressed as

$$x_{10} - \mu \cdot g/x_2 < 0, \tag{5.30a}$$
$$-x_{10} - \mu \cdot g/x_2 < 0, \tag{5.30b}$$
$$x_9 - l_r \cdot x_{10}/x_2 - \bar{\alpha}_r < 0, \tag{5.30c}$$
$$-x_9 + l_r \cdot x_{10}/x_2 - \bar{\alpha}_r < 0, \tag{5.30d}$$

where $\bar{\alpha}_r$ stands for the upper limit for the rear tire sideslip angle.

Finally, the constraints of initial and terminal states, along with the state and input limits, must be enforced. The planned path starts at the initial state $\boldsymbol{x}_{t=0} = \boldsymbol{x}_{\text{init}}$ and ends at specific terminal states defined as $\boldsymbol{x}^i_{t=t_f} = \boldsymbol{x}^i_{\text{end}}$, wherein t_f represents the planning horizon and the superscript i is the indices of particular states with terminal constraints.

Thus, this section formulates the local trajectory optimization problem as a constrained nonlinear optimization problem with multiple objectives and multiple constraints. The optimization problem is expressed as follows.

$$\begin{aligned} &\min_{\boldsymbol{u}_p, \boldsymbol{u}_v} \int_0^{t_f} (\Xi_r + \Xi_v + \Xi_p)\mathrm{d}t + J_c \\ &\text{s.t.} \quad \dot{\boldsymbol{x}} = f(\boldsymbol{x}, \boldsymbol{u}_p, \boldsymbol{u}_v),\ h_v(\boldsymbol{x}, \boldsymbol{u}_v) = 0, \\ &\psi_c(\boldsymbol{x}) > 1, \\ &\psi_t(\boldsymbol{x}) < 0,\ \psi_l(\boldsymbol{x}) < 0, \\ &\boldsymbol{x}_{t=0} = \boldsymbol{x}_{\text{init}},\ \boldsymbol{x}^i_{t=t_f} = \boldsymbol{x}^i_{\text{end}}, \\ &\underline{x} \le \boldsymbol{x} \le \bar{\boldsymbol{x}},\ \underline{u}_v \le \boldsymbol{u}_v \le \bar{\boldsymbol{u}}_v, \end{aligned} \tag{5.31}$$

where $\underline{x}$ and $\bar{\boldsymbol{x}}$ are the upper and lower bounds of the state variables, respectively, and similar notations are used for $\underline{u}_v$ and $\bar{\boldsymbol{u}}_v$. These bounds are essential for ensuring both safety and feasibility in trajectory planning by preventing states or inputs from violating physical limitations or operational constraints.

The local path planning task is solved by the nonlinear optimization solver provided by CasADi, which aims to minimize the cost function while satisfying all the constraints. CasADi is an open-source software framework for the nonlinear optimization problems [2]. In particular, the IPOPT solver is utilized to solve the formulated path-planning problem. IPOPT is a robust and widely used interior-point algorithm for large-scale nonlinear optimization.

5.4.3 EXPERIMENTAL RESULTS OF PATH PLANNING

To validate the effectiveness of the proposed path planning method, this subsection conducts a series of experiments, for example, the scenario of lane-changing, scenario of overtaking slower cars ahead, scenario of collision avoidance. All these scenarios are implemented on both straight and curved roads. For comparison, we also utilize the widely recognized sampling-based trajectory planning method [139] as a benchmark, which is referred to as the common method in this chapter. Unlike the proposed method, this baseline method does not consider chassis domain dynamics when performing the path planning task. Through these test scenarios, this subsection aims to assess the robustness, smoothness, and safety performance of the proposed approach.

5.4.3.1 Lane-Change on Straight Roads

This experiment carries out the lane-change maneuver on straight roads (LC-S), which is a typical scenario in urban driving. Fig. 5.20 presents the planned lane-change trajectories for both the proposed method and the common method. In this section, we utilize a color map to denote vehicle velocities in the demonstrations of all test trajectories, as illustrated in Fig. 5.20. In this lane-change experiment, the ego vehicle operates at an initial speed of 30 km/h, and its driving speed keeps almost constant during this lane-change maneuver. From approximately 2 s to 6 s, the ego car performs the lane-change maneuver, during which the vehicle runs about 30 m. The common method performs a sharper lane-change path, while the proposed method generates a smoother trajectory. It is observed that the peak steering angle performed by the proposed method is approximately -4° at 2.5 s. Besides, the peak values of the vehicle's sideslip angle and yaw rate achieved by the proposed method are considerably smaller than those obtained by the common method. The phenomenon implies that the common method performs a less stable lane-change maneuver. The proposed method can provide a wider dynamic stability margin by considering chassis domain dynamics performance in the trajectory planning procedure. It can be found that the proposed method outperforms the common method, and the proposed method can generate a smoother trajectory with better dynamic performance.

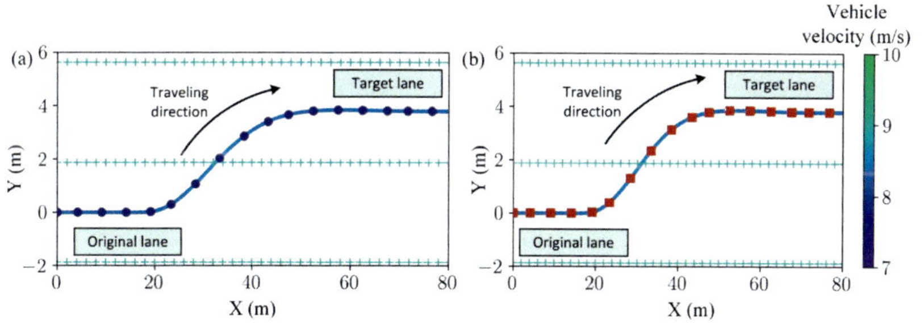

Figure 5.20 The planned lane-change trajectories on straight roads, (a) the proposed method, (b) the common method.

5.4.3.2 Static Collision Avoidance on Straight Roads

This experiment conducts the static collision avoidance maneuver on straight roads (SA-S), and the results are illustrated in Fig. 5.21. In this case, a stationary obstacle vehicle is positioned 50 m ahead of the ego car. The ego car performs steering operations to avoid the obstacle when the distance between the two cars is approximately 20 m. The obstacle-avoidance process spans a duration of approximately 3 s to 11.5 s. From Fig. 5.21, it is evident that the ego car controlled by the common method tends to perform a sharper steering maneuver to return to the original lane after avoiding the obstacle. However, this maneuver introduces a heightened risk from the standpoint of both collision avoidance effectiveness and chassis dynamic stability. In terms of vehicle lateral dynamics, the proposed method performs better performance in the SA-S test. The maximum yaw rate achieved by the proposed method is 15°/s, compared to 18°/s achieved by the common method. Similarly, the peak sideslip angle is reduced to 3.2° by the proposed method, as opposed to 4° with the common method. The results of the SA-S test indicate that the proposed method enhances the stability of vehicle chassis dynamics when performing the collision-avoidance maneuver. This test case highlights the effectiveness of the proposed chassis-domain-aware trajectory planning method. By integrating collision avoidance with chassis dynamic performance, the proposed method can guarantee a larger stability margin while successfully avoiding stationary obstacles ahead.

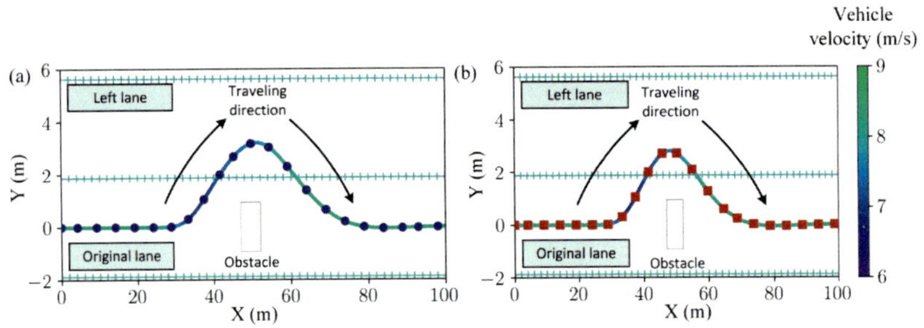

Figure 5.21 The planned collision-avoidance trajectories on straight roads, (a) the proposed method, (b) the common method.

5.4.3.3 Overtaking on Straight Roads

This experiment conducts the overtaking maneuver on straight roads (OT-S). Fig. 5.22 illustrates the planned overtaking trajectories. In this case, the ego vehicle encounters a slow-moving car traveling at a speed of 10 km/h. An overtaking maneuver is performed to pass the slow-moving car as the ego car approaches the preceding car. It can be observed that there are trajectory overlaps in Fig. 5.22. However, it is necessary to clarify that these overlaps do not imply collisions between the two cars. The trajectory figure does not convey the temporal information of the overtaking process. The overlapping positions demonstrated in Fig. 5.22 correspond to different time instances.

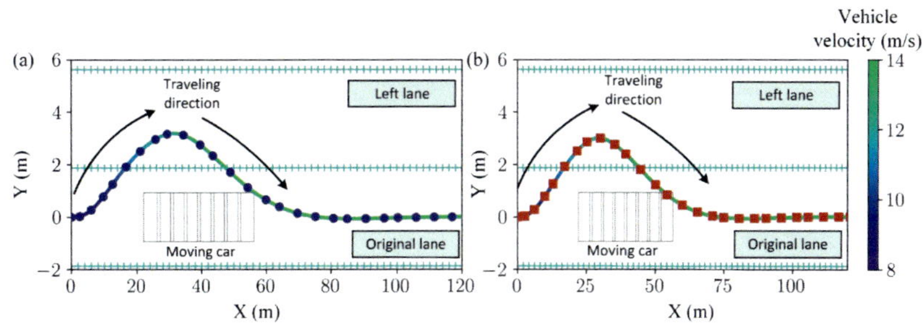

Figure 5.22 Overtaking trajectory profiles in the straight road case, (a) the proposed method, (b) the common method.

In this test scenario, the preceding car runs along the road from 25 m to 55 m during the test duration of 10.8 s. Meanwhile, the ego car initiates the overtaking maneuver with its speed increasing gradually from 30 km/h to 50 km/h. Then, the ego car runs at a constant speed after completing the overtaking process. The longitudinal acceleration controlled by the proposed method reaches a maximum of 1.791 m/s^2 at 1.5 s, which is smaller than the peak value of 2.020 m/s^2 controlled by the common method. Moreover, the common method performs the acceleration operation earlier compared to the proposed method.

In terms of vehicle lateral motion, the ego car performs a lane-change operation to the left lane to overtake the slow-moving car, and it returns to the original lane after successfully passing the preceding car. The maximum steering angle required for the lane change to the left lane is about -4° at 0.5s. For the return to the original lane, the maximum steering angle is reduced to approximately 2.6° at 3s, which is smaller than the former value. This phenomenon is attributed to the fact that the ego vehicle's speed increases during the overtaking process. Thus, the steering angles for returning to the original lane, which are performed at a higher speed, are smaller, so that chassis domain dynamic performance can be guaranteed. In contrast, all corresponding peak values controlled by the common method are slightly larger. This notable difference can be attributed to the superior trajectory optimization capabilities of our proposed method, which considers vehicle chassis dynamic responses during the overtaking procedure. These findings further validate the performance of the proposed method in planning smooth, efficient, and dynamic optimal trajectories for the overtaking maneuver.

5.4.3.4 Lane-Change on Curvy Roads

This experiment conducts the lane-change maneuver on curvy roads (LC-C). As shown in Fig. 5.23, the ego car performs a lane-change operation on a curvy road with a road curvature of 0.0067 m^{-1}. The LC-C test begins with the same initial speed as the LC-S test. It can be observed that the ego car

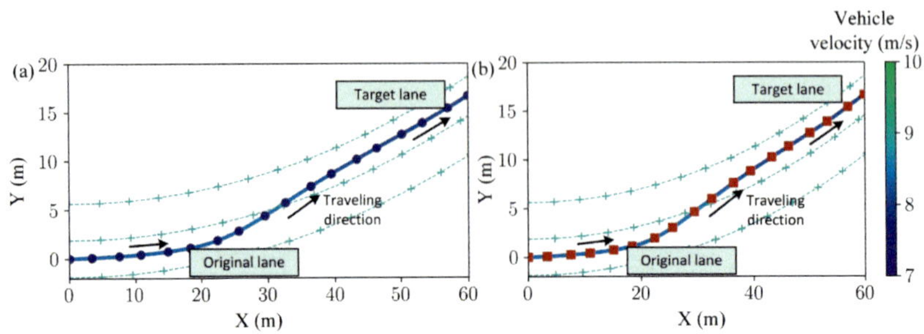

Figure 5.23 Lane-change trajectory profiles in the curvy road case, (a) the proposed method, (b) the common method.

achieves a similar and reasonable outcome to that achieved in the LC-S test, which suggests the performance consistency of the proposed method. The ego car continues to drive on the curved road with a steering angle of about 1.5° after performing the lane-change maneuver. Vehicle yaw rate is about 3.2°/s during this process.

Notably, the proposed method can perform the lane-change maneuver on curvy roads more smoothly compared to the common method. The maximum value of vehicle lateral speed controlled by the common method is larger than that controlled by the proposed method during the lane-change maneuver. This result validates the robustness and effectiveness of the proposed method. The proposed method can effectively balance the performance of chassis dynamics and lane-change trajectory on curvy roads.

5.4.3.5 Static Collision Avoidance on Curvy Roads

This experiment conducts the static collision avoidance maneuver on curvy roads (SA-C). The planned trajectories of the SA-C test are illustrated in Fig. 5.24. This collision avoidance maneuver on curvy roads begins at approximately 3s and ends at about 10s. As observed in the SA-C test, the collision-avoidance trajectory generated by the common method tends to be sharper than that generated by the proposed method, which is similar to the results of the SA-S test. The peak values of vehicle lateral dynamics states controlled by the proposed method are observed to be nearly identical to those controlled by the common method in this scenario. The underlying reason for this similarity can be attributed to the greater deceleration controlled by the common method. In contrast, the proposed method achieves smoother maneuvers with optimized control, which validates its ability to balance chassis dynamic performance and collision avoidance effectively.

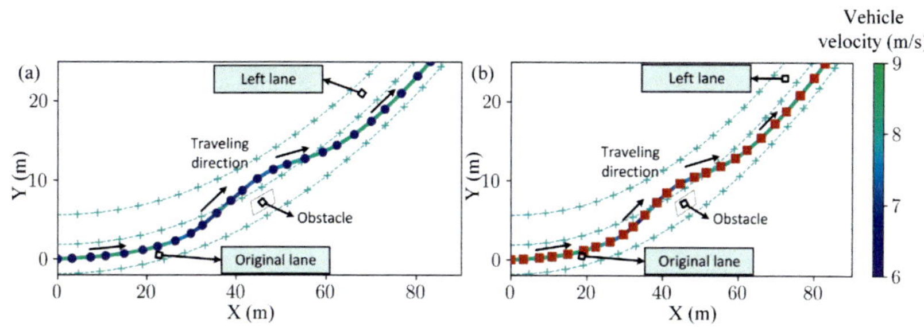

Figure 5.24 Collision avoidance trajectory profiles in the curvy road case, (a) the proposed method, (b) the common method.

5.4.3.6 Overtaking on Curvy Roads

This experiment conducts the overtaking maneuver on curvy roads (OT-C). The planned overtaking trajectories on curvy roads are illustrated in Fig. 5.25. In this test scenario, the ego car overtakes a slow-moving car on a curvy road, which highlights the capability of the proposed method to perform coordinated longitudinal and lateral driving effectively. The proposed method performs the acceleration maneuver approximately 2s later than the common method. The maximum values of lateral dynamic states achieved by the proposed method are smaller than those achieved by the common method. This demonstrates the superior dynamic performance achieved by the proposed method, since it minimizes abrupt changes in vehicle dynamic states during the overtaking maneuver.

The summarized results of all test scenarios are presented in Fig. 5.26. This figure demonstrates the peak magnitudes of various vehicle dynamic states achieved by the two methods, which can highlight a comprehensive comparison across the different test scenarios.

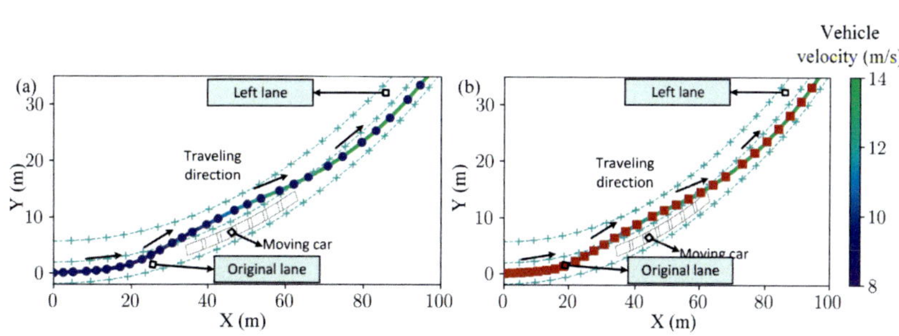

Figure 5.25 Overtaking trajectory profiles in the curvy road case, (a) the proposed method, (b) the common method.

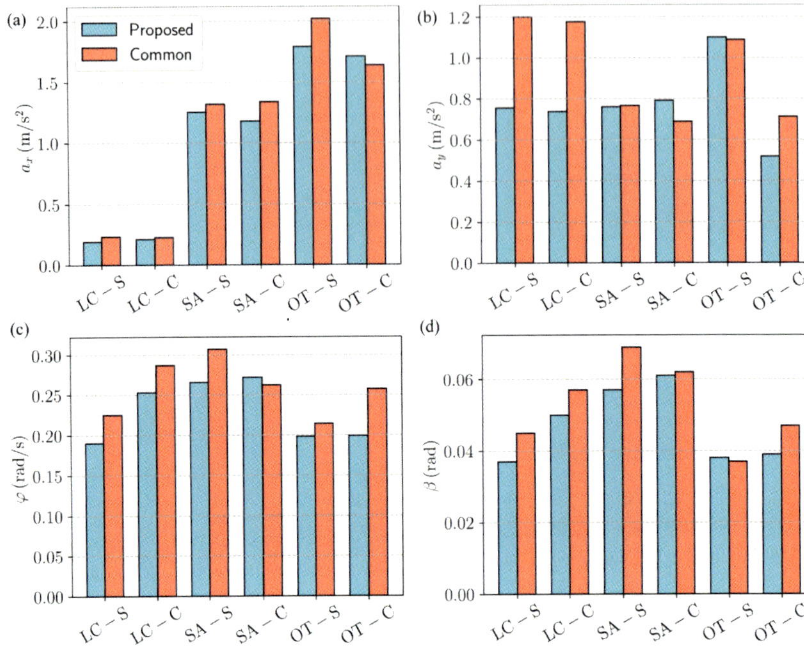

Figure 5.26 Summary of peak values of key dynamic states in various experiments.

5.4.4 SUMMARY

This section introduces a chassis domain dynamics-aware trajectory planning method to address the optimal local trajectory planning problem with consideration of vehicle dynamics, which can balance geometric safety, dynamic stability, and path smoothness in various scenarios. Various test scenarios, including lane change, static obstacle avoidance, and overtaking maneuvers on both straight roads and curvy roads, are carried out. Those results are presented and discussed, which show the effectiveness and feasibility of our proposed method. These results highlight the proposed method's strong adaptability to various traffic scenes.

5.5 CONCLUSION

A chassis domain dynamics-aware trajectory planning scheme is proposed to address the optimal local trajectory planning problem while accounting for comprehensive chassis domain dynamics. By incorporating global chassis dynamics into the path planning process, this method ensures that chassis dynamics performance can be effectively integrated into local trajectory optimization. First, global chassis dynamics is investigated to let essential dynamics performance be sufficiently involved in local trajectory planning. We develop a quantitative evaluation metric for AVs' chassis dynamics performance, including dynamics stability and riding comfort in a rigorous mathematical way. Then, to cope with complicated and dynamic traffic environments together with the strong nonlinearity of chassis dynamics, this chassis-domain-aware trajectory planning method contains geometrical smoothness, chassis dynamics, and feasible safety in the optimization of trajectory planning. Thus, a multi-objective optimization problem of trajectory planning is formulated to provide proper local trajectories. Various test scenarios including lane-change, static obstacle avoidance, and overtaking maneuvers on both straight roads and curvy roads are carried out. Those results show the effectiveness and feasibility of our proposed scheme and indicate our methods' strong adaptability to various traffic scenes. These findings show that the global chassis dynamics-oriented trajectory planning method merits further investigation to speed up the mass deployment of AVs.

6 Multi-Actuator Integrated Control for Chassis Domain Dynamics Optimization

This chapter focuses on the coordinated control for autonomous driving, which involves multiple chassis actuators and multiple control objectives. Dynamic stability and trajectory tracking are two essential control objectives for AVs. First, considering vehicle time-varying and uncertain parameters, this chapter introduces a linear matrix inequality-based coordinated control method for the integrated control of active front steering and differential yaw control. Typical experiments are carried out to verify the effectiveness of the proposed coordinated control method. Then, this chapter proposes a global chassis multi-actuator integrated control framework to realize the optimization of chassis global dynamics, including trajectory tracking and handling stability. In this framework, each wheel corner is fully adjustable and controllable, and the global dynamics are constructed according to the analysis of distributed and controllable wheel modules and vehicle dynamics. The proposed framework is based on a game-theoretical control method, and the chassis global dynamics can be optimized by achieving a Nash equilibrium. Experimental results also illustrate the feasibility and effectiveness of the proposed control method.

6.1 BACKGROUND

As we all know, chassis execution performs the final step in translating high-level control commands into physical actions, which fundamentally depends on the tire forces of four wheels. Therefore, robust and accurate chassis execution is essential to guarantee the safe and reliable operation of AVs. Chassis dynamic stability is a critical issue in the context of AVs, and it is required to ensure vehicle safety and perform vehicle motion control in various driving conditions. Besides, path tracking is another essential task for AVs, which controls the vehicle to follow the expected trajectories at specific speeds. Both chassis dynamic stability and path tracking highlight the significance of advanced chassis control technologies for AVs.

Vehicle chassis has become a complicated mechatronic system with the rapid development of automotive electronic control technologies, which involves multiple specialized actuators. Thus, the coordinated control of these actuators is both a critical challenge and a compelling subject in the field of automotive engineering. Considerable efforts have been devoted to enhancing vehicle dynamic performance, path-tracking performance, and

DOI: 10.1201/9781003583929-6

driving efficiency via the integrated control of multiple chassis actuators [100, 74, 25].

How to improve the performance of vehicle dynamic stability has been investigated for decades. Van Zanten, at Bosch Corporation, introduced the integration of direct yaw moment control (DYC) and vehicle slip ratio control [129]. DYC usually performs an additional yaw moment via vehicle differential braking to improve vehicle handling and stability, particularly in critical driving conditions, such as sharp turns or slippery surfaces. On the other hand, active front steering (AFS) controls the vehicle steering system by introducing an auxiliary steering angle to improve vehicle dynamic responses [154]. A variety of control methods have been investigated to coordinate AFS and DYC to enhance vehicle dynamic performance, such as the robust gain-scheduling approach [150], and the model predictive control method [108]. Our previous study proposed a robust H_∞ coordinated controller for vehicle dynamic control via DYC and AFS [26], which is derived based on the linear matrix inequality (LMI) technique.

Existing works on path tracking mainly focus on performing lateral trajectory tracking by calculating corresponding steering angles or controlling vehicle longitudinal motions by calculating throttle or brake pedal openings [10, 155]. In the context of autonomous driving, the performance of both trajectory tracking and vehicle dynamic stability must be considered simultaneously. In this regard, four-wheel independent drive vehicles (4WIDVs) have been developed to enhance vehicle maneuverability, flexibility, and high-efficiency transmission, for example, the active distribution of the driving torques of all wheels can output additional yaw moments to improve chassis dynamic responses. Besides, four-wheel steering (4WS) was also investigated to improve vehicle maneuverability and stability [45, 22]. In addition, some studies focused on the coordinated control of 4WIDVs and investigated various approaches for the coordinated control of the braking system, the steering system, and the four-wheel-independent-drive traction system [21, 19, 57]. In the aspect of the distribution of motor torques, various control strategies have been introduced, for example, the optimal torque energy-saving allocation method [72], the torque vectoring algorithm to enhance the steering performance [107], and the multi-objective optimization of four-wheel torque distribution using a particle swarm algorithm [134]. Additional works have focused on taking advantage of the 4WID system and integrating it with other chassis components [48, 20], which merits further investigations.

In light of the growing deployment of embedded control technologies in the automotive industry, some investigations have been conducted to design integrated control architectures to efficiently manage multiple chassis subsystems [93]. A global chassis control approach, which utilized the inverse vehicle modeling method, was introduced in the literature [3]. A unified and reconfigurable vehicle dynamics model and a corresponding coordinated control strategy were developed in the literature [6]. In addition, a hybrid adaptive chassis control framework was proposed by the literature [69] These studies provide valuable insights into the development of integrated control architectures for vehicle chassis systems.

6.2 COORDINATED CONTROL OF AFS AND DYC

This section introduces an H_∞ feedback control method for enhancing vehicle dynamic stability, which utilizes the LMI theory. First, a lateral dynamic model is established based on the polytopic uncertainty method, which explicitly considers the time-varying speed and vehicle mass. All feasible uncertainty parameter values are covered by a polytope, which consists of finite vertices. Second, a robust controller is derived to coordinate the AFS and DYC. Then, the proposed method performs accurate allocation of braking pressures of all-wheel cylinders to achieve the required yaw moment, and the auxiliary steering angle of AFS is also adjusted based on vehicle dynamic states.

Fig. 6.1 illustrates the architecture of the proposed integrated H_∞ controller, which consists of two layers, the upper H_∞ control layer and the lower distribution control layer. The upper layer performs the coordinated control of AFS and DYC and generates the desired additional yaw moment. The lower layer apportions the required yaw moment to the brake pressure of each wheel cylinder and the auxiliary steering angle of AFS.

6.2.1 POLYTOPIC MODELING

Based on vehicle dynamic modeling introduced in Section 2.4, this subsection establishes a vehicle lateral dynamic model with polytopic uncertainty. A polytope consists of a set of vertices, and it can represent the entire range of

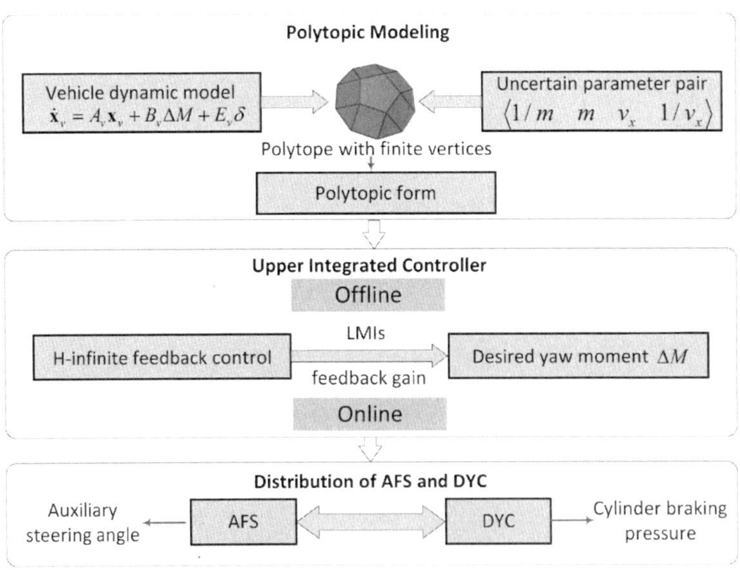

Figure 6.1 The architecture of the proposed integrated H_∞ controller.

parameter variations. This subsection uses this method to model parameter uncertainties including vehicle longitudinal velocity v_x and mass $m(I_z)$.

Given the definition of the state vector $\mathbf{x}_v = [\beta, \gamma]^T$ and control input ΔM, this subsection formulates the vehicle lateral dynamic model for the design of the H_∞ feedback controller, as shown in Eq. (6.1).

$$\dot{\mathbf{x}}_v = A_v \mathbf{x}_v + B_v \Delta M + E_v \delta, \tag{6.1}$$

with

$$A_v = \begin{bmatrix} \dfrac{k_f + k_r}{m v_x} & \dfrac{l_f k_f - l_r k_r}{m v_x^2} - 1 \\ \dfrac{l_f k_f - l_r k_r}{I_z v_x} & \dfrac{l_f^2 k_f + l_r^2 k_r}{I_z v_x} \end{bmatrix}, \tag{6.2}$$

$$B_v = \begin{bmatrix} 0 \\ 1/I_z \end{bmatrix}, E_v = \begin{bmatrix} -\dfrac{k_f}{m v_x} \\ -\dfrac{l_f k_f}{I_z} \end{bmatrix}. \tag{6.3}$$

The principle of vehicle dynamic stability lies in the accurate tracking of the reference values for both vehicle sideslip angle and yaw rate. The reference values $\mathbf{r} = [\beta_n, \gamma_n]^T$ are defined as shown in Eq. (6.4). Then, we extend the system model in Eq. (6.1) by introducing augmented states, and the augmented state vector is defined by Eq. (6.5).

$$\mathbf{r} = \begin{bmatrix} \dfrac{v_x/l}{1 + K v_x^2} & \dfrac{l_r + m l_f v_x^2/(k_r l)}{l(1 + K v_x^2)} \end{bmatrix}^T \delta = R\delta. \tag{6.4}$$

$$\mathbf{x} = \begin{bmatrix} \mathbf{x}_v & \int_0^t (\mathbf{r} - \mathbf{x}_v) dt \end{bmatrix}^T. \tag{6.5}$$

Based on the definition of the augmented state vector, the augmented system model is formulated as follows.

$$\dot{\mathbf{x}} = A\mathbf{x} + B\Delta M + E\delta, \tag{6.6}$$

with

$$A = \begin{bmatrix} A_v & 0 \\ -I & 0 \end{bmatrix}, B = \begin{bmatrix} B_v \\ 0 \end{bmatrix}, E = \begin{bmatrix} E_v \\ R \end{bmatrix}. \tag{6.7}$$

In this section, vehicle longitudinal speed and vehicle mass are constrained within specific bounds, which are defined as $[m_{\min}, m_{\max}]$ and $[v_{x,\min}, v_{x,\max}]$. The system model involves parameters v_x, m, $1/v_x$, and $1/m$. To account for the variations in these parameters, we design a polytope comprising 16 vertices that cover all potential combinations of the parameter pair $\langle v_x, m, 1/v_x, 1/m \rangle$. Thus, the parameter pair can be represented by the following expression:

$$\begin{cases} 1/m = \sum\limits_{j=1}^{2} \tilde{\vartheta}_j \tilde{\xi}_j, m = \sum\limits_{j=1}^{2} \widehat{\vartheta}_j \widehat{\xi}_j, \\ v_x = \sum\limits_{j=1}^{2} \tilde{\psi}_j \tilde{\zeta}_j, 1/v_x = \sum\limits_{j=1}^{2} \widehat{\psi}_j \widehat{\zeta}_j, \end{cases} \tag{6.8}$$

where

$$\begin{cases} \tilde{\xi}_1 = 1/m_{\max}, \tilde{\xi}_2 = 1/m_{\min}, \widehat{\xi}_1 = m_{\min}, \widehat{\xi}_2 = m_{\max}, \\ \tilde{\zeta}_1 = v_{x,\min}, \tilde{\zeta}_2 = v_{x,\max}, \widehat{\zeta}_1 = 1/v_{x,\max}, \widehat{\zeta}_2 = 1/v_{x,\min}. \end{cases} \tag{6.9}$$

$$\begin{cases} \tilde{\vartheta}_1 = \dfrac{\tilde{\xi}_2 - 1/m}{\tilde{\xi}_2 - \tilde{\xi}_1}, \tilde{\vartheta}_2 = \dfrac{1/m - \tilde{\xi}_1}{\tilde{\xi}_2 - \tilde{\xi}_1}, \\ \widehat{\vartheta}_1 = \dfrac{\widehat{\xi}_2 - m}{\widehat{\xi}_2 - \widehat{\xi}_1}, \widehat{\vartheta}_2 = \dfrac{m - \widehat{\xi}_1}{\widehat{\xi}_2 - \widehat{\xi}_1}, \\ \tilde{\psi}_1 = \dfrac{\tilde{\zeta}_2 - v_x}{\tilde{\zeta}_2 - \tilde{\zeta}_1}, \tilde{\psi}_2 = \dfrac{v_x - \tilde{\zeta}_1}{\tilde{\zeta}_2 - \tilde{\zeta}_1}, \\ \widehat{\psi}_1 = \dfrac{\widehat{\zeta}_2 - 1/v_x}{\widehat{\zeta}_2 - \widehat{\zeta}_1}, \widehat{\psi}_2 = \dfrac{1/v_x - \widehat{\zeta}_1}{\widehat{\zeta}_2 - \widehat{\zeta}_1}. \end{cases} \tag{6.10}$$

Based on the above definitions, we can reformulate the system model in its polytopic form, which represents the varying parameters by utilizing their defined bounds:

$$\begin{aligned} \dot{\mathbf{x}}(k+1) &= \sum_{j=1}^{16} \rho_j(k)\left[A_j \mathbf{x}(k) + B_j u(k) + E_j \delta(k)\right] \\ &= A(\rho)\mathbf{x}(k) + B(\rho)u(k) + E(\rho)\delta(k), \end{aligned} \tag{6.11}$$

where

$$\begin{cases} \rho = \begin{bmatrix} \rho_1 & \rho_2 & \cdots & \rho_{16} \end{bmatrix}^T, \\ \rho_1 = \bar{\rho}_{1111}, \rho_2 = \bar{\rho}_{1112}, \rho_3 = \bar{\rho}_{1121}, \cdots \rho_{16} = \bar{\rho}_{2222}, \\ \bar{\rho}_{klhn} = \tilde{\vartheta}_k \widehat{\vartheta}_l \tilde{\psi}_h \widehat{\psi}_n. \end{cases} \tag{6.12}$$

It is worth noting that the system matrices A_j, B_j, and E_j are inherently time-invariant, and this property could facilitate the derivation of the H_∞ controller.

Furthermore, based on the analysis of the expressions in Eq. (6.10) and Eq. (6.12), we can yield the following expression:

$$\sum_{j=1}^{16} \rho_j(t) = 1, \rho_j(t) \geq 0. \tag{6.13}$$

6.2.2 DESIGN OF COORDINATED CONTROLLER

In this subsection, we introduce the design of the H_∞ coordinated controller. The polytopic representation of the system model is the foundation for the design of the H_∞ controller. The main objective of the upper layer is to generate the additional yaw moment, which is required to track the reference values of vehicle sideslip angle and yaw rate.

To provide a comprehensive explanation of the controller design, we first introduce some basic definitions and lemmas.

Definition 6.1 *A matrix polytope can be described as a convex hull, which is formed by a set of matrices of identical dimensions [4]. It can be expressed as follows.*

$$Co\{M_i, i=1,2,...,r\} := \left\{ \sum_{i=1}^{r} \alpha_i M_i : \alpha_i \geq 0, \sum_{i=1}^{r} \alpha_i = 1 \right\}. \tag{6.14}$$

Then, given a linear parameter-varying system which is expressed by $\langle A'(\upsilon), B'(\upsilon), C'(\upsilon), D'(\upsilon) \rangle$, *we make the following assumptions:*

1) The dependence of the state-space matrices on the parameter υ *is affine. In other words, the matrices vary linearly with respect to* υ.

2) The time-varying parameter υ *is constrained within a polytope* Υ, *defined by its vertices* $\omega_1, \omega_2, \ldots, \omega_r$, *which is presented as follows.*

$$\upsilon \in \Upsilon := Co\{\omega_1, \omega_1, ..., \omega_r\}. \tag{6.15}$$

Based on these conditions, we can reformulate the linear parameter-varying system as

$$\begin{pmatrix} A'(\upsilon) & B'(\upsilon) \\ C'(\upsilon) & D'(\upsilon) \end{pmatrix} \in \mho \tag{6.16}$$

with

$$\mho := Co\left\{ \begin{pmatrix} A'_i & B'_i \\ C'_i & D'_i \end{pmatrix} := \begin{pmatrix} A'(\omega_i) & B'(\omega_i) \\ C'(\omega_i) & D'(\omega_i) \end{pmatrix} \right\}, i=1,...,r. \tag{6.17}$$

Definition 6.2 *Assume we have a real symmetric matrix* L_D, *and a real matrix* R_1. *They are required to meet the following condition:*

$$D = \left\{ s \in C : L_D + sR_1 + \bar{s}R_1^T < 0 \right\}, \tag{6.18}$$

where D *stands for an LMI region, and* C *represents the complex set.*

Lemma 6.1 *If and only if there is a symmetric positive-definite matrix* X_D *such that Eq. (6.19) holds, we state that all eigenvalues of* R_2 *are in the LMI region* D, *in other words, the real matrix* R_2 *is* D*-stable [90].*

$$M_D(R_2, X_D) = L_D \otimes X_D + R_1 \otimes (X_D R_2) + R_1^T \otimes (R_2^T X_D) \tag{6.19}$$

Specifically, for the D-stability of a disk LMI region (q, r), *the following condition must be satisfied:*

$$\begin{bmatrix} -rX_D & qX_D + X_D R_2 \\ * & -rX_D \end{bmatrix} < 0, X_D > 0. \tag{6.20}$$

Lemma 6.2 *Given a transfer function (not necessarily minimal) for a continuous or discrete-time plant, which is represented by* $G(s) = D'' + C''(sI - A'')^{-1}B''$*, the following two statements are equivalent:*

1) A positive definite matrix X *exists such that the following expression holds:*

$$\begin{bmatrix} sys\{A^T X\} & XB & C^T \\ * & -\gamma_H I & D^T \\ * & * & -\gamma_H I \end{bmatrix} < 0. \tag{6.21}$$

Here, $sys\{\odot\}$ *stands for* $\odot + \odot^T$*, and the symbol* $*$ *represents a symmetric matrix's symmetric elements.*

2) The stable system plant A'' *satisfies the following condition:*

$$\left\| D'' + C''(sI - A'')^{-1}B'' \right\|_\infty < \gamma_H. \tag{6.22}$$

Lemma 6.3 *The following three statements about the polytopic system as shown in Eq. (6.16) are equivalent.*

1) The system plant is stable and its H_∞ *performance index is bounded by* γ_H.

2) There is a matrix $P > 0$ *such that the condition as expressed by Eq. (6.24) holds for all system matrices*

$$\begin{pmatrix} A'(\upsilon) & B'(\upsilon) \\ C'(\upsilon) & D'(\upsilon) \end{pmatrix} \in \mho \tag{6.23}$$

$$\begin{bmatrix} sys\left\{A'^T(\upsilon)P\right\} & PB'(\upsilon) & C'^T(\upsilon) \\ * & -\gamma_H I & 0 \\ * & * & -\gamma_H I \end{bmatrix} < 0, \tag{6.24}$$

3) There is a matrix $P > 0$ *that satisfies the following set of LMIs:*

$$\begin{bmatrix} sys\left\{A_i'^T P\right\} & PB'_i & C_i'^T \\ * & -\gamma_H I & 0 \\ * & * & -\gamma_H I \end{bmatrix} < 0, i = 1, 2, \ldots, r. \tag{6.25}$$

Based on the above definitions and analysis, we introduce a theorem as follows [26].

Theorem 6.1 *For a linear time-invariant system, which is presented in Eq. (6.26), a state feedback control law* $u_I = K_I x_I$ *is designed to stabilize the system and guarantee that the condition* $\|z_I\|_2 / \|w_I\|_2 < \gamma_I$ *holds. This is achievable if and only if there exists a positive definite symmetric matrix* X_I *and a matrix* L_I *such that the statement as presented in Eq. (6.27) holds. The feedback gain matrix can be calculated by* $K_I = L_I X_I^{-1}$.

$$\begin{cases} \dot{x}_I = A_I x_I + B_{I,1} w_I + B_{I,2} u_I, \\ z_I = C_{I,1} x_I + D_{I,11} w_I + D_{I,12} u_I, \\ y_I = C_{I,2} x_I + D_{I,21} w_I + D_{I,22} u_I, \end{cases} \tag{6.26}$$

$$\begin{bmatrix} \Lambda_1 & B_{I,1} & \wp_1 \\ * & -\gamma_I I & D_{I,11}^T \\ * & * & -\gamma_I I \end{bmatrix} < 0, \tag{6.27}$$

where $\Lambda_1 = sys\{A_I X_I + B_{I,2} L_I\}$, *and* $\wp_1 = X_I C_{I,1}^T + L_I^T D_{I,12}^T$.

Thus, based on theorem 6.1, the corresponding state feedback control law can be designed for each vertex of the system plant expressed by Eq. (6.16). The overall control law for the controlled system can be yielded based on the control law for each vertex. Accordingly, this chapter develops the control law, which is expressed as $\Delta M = K_v \mathbf{x}$ for vehicle lateral dynamic system given by Eq. (6.29). K_v is derived by calculating a combination of K_j, which is the control gain of the $j-th$ system vertex:

$$K_v = \sum_{j=1}^{16} \rho_j(k) K_j. \tag{6.28}$$

$$\begin{cases} \dot{\mathbf{x}}(k+1) = \sum\limits_{j=1}^{16} \rho_j(k)\left[A_j \mathbf{x}(k) + B_j u(k) + E_j \delta(k)\right], \\ \mathbf{z}(k+1) = C\mathbf{x}(k+1), C = \begin{bmatrix} 0 & 0 & 1 & 1 \end{bmatrix}. \end{cases} \tag{6.29}$$

Furthermore, to design the control gain K_j for the $j-th$ vertex, we introduce a symmetric positive-definite matrix X_v and a matrix, which is defined as

$$L_j = K_j X_v, j = 1, 2, ..., 16 \tag{6.30}$$

K_j can be derived by the minimization of a performance index γ_o, which is formulated as follows.

$$\begin{aligned} &\min_{X_v, L_j} \gamma_o \\ &s.t. \\ &\begin{bmatrix} -rX_v & qX_v + A_j X_v + B_j L_j \\ * & -rX_v \end{bmatrix} < 0, \\ &\begin{bmatrix} \Lambda & E_j & \wp \\ * & -\gamma_o I & 0 \\ * & * & -\gamma_o I \end{bmatrix} < 0, \end{aligned} \tag{6.31}$$

where $\Lambda = sys\{A_j X_v + B_j L_j\}, \wp = X_v C^T$.

6.2.3 DISTRIBUTION OF ADDITIONAL YAW MOMENT

The desired additional yaw moment, required by the upper layer, is distributed based on vehicle real-time dynamic states. In the architecture of the proposed integrated H_∞ controller, AFS and DYC perform in concert with each other to enhance vehicle dynamic performance. It is essential to establish an allocation strategy between AFS and DYC. First, this subsection introduces the differences in operation characteristics between AFS and DYC. When the vehicle operates with small tire sideslip angles, vehicle tires work within their linear regions. This suggests that vehicle steering wheels still have large controllable steering margins, and AFS can perform yaw rate control effectively. In this case, the steering adjustments provided by AFS are sufficient to control vehicle yaw moment. On the other hand, if vehicle tires operate in corresponding nonlinear regions, the effectiveness of AFS may decline since vehicle steering tires work with large sideslip angles in this situation. While DYC can perform differential braking to generate the required yaw moment.

This subsection designs a factor $\mathfrak{A}$ to represent vehicle dynamic status, which is derived using vehicle sideslip angle and lateral acceleration. It can be expressed as:

$$\mathfrak{A} = \sqrt{g_1 a_y^2 + g_2 \beta^2}, \tag{6.32}$$

where β is vehicle sideslip angle, a_y represents vehicle lateral acceleration. The coefficients g_1, and g_2 are scaling factors to balance the corresponding contributions of a_y and β.

Thus, this chapter develops a dynamic weighting function to manage the yaw moment allocation of AFS and DYC based on the designed factor $\mathfrak{A}$. Specifically, the weighting function of DYC is formulated as shown in Eq. (6.33). Then, the desired yaw moment is performed by the combination of $q_{DYC} \cdot \Delta M$ via DYC and $1 - q_{DYC} \cdot \Delta M$ via AFS.

$$q_{DYC} = \begin{cases} 0, 0 \le \mathfrak{A} < \mathfrak{A}_{\text{low}} , \\ \dfrac{2}{(\mathfrak{A}_{\text{upper}} - \mathfrak{A}_{\text{lowv}})^2} (\mathfrak{A} - \mathfrak{A}_{\text{low}})^2 , \mathfrak{A}_{\text{low}} \le \mathfrak{A} < \mathfrak{A}_{\text{middle}} , \\ 1 - 2\left(1 - \dfrac{\mathfrak{A} - \mathfrak{A}_{\text{low}}}{\mathfrak{A}_{\text{upper}} - \mathfrak{A}_{\text{low}}}\right)^2 , \mathfrak{A}_{\text{middle}} \le \mathfrak{A} < \mathfrak{A}_{\text{upper}} , \\ 1, \mathfrak{A}_{\text{upper}} \le \mathfrak{A} < \mathfrak{A}_{\text{end}} , \end{cases} \tag{6.33}$$

where $\mathfrak{A}_{low}$, $\mathfrak{A}_{middle}$, and $\mathfrak{A}_{upper}$ denote predefined threshold values, which are calibrated based on the analysis of experimental results. These thresholds segment the spectrum of the yaw moment allocation and guarantee a seamless transition in the control effort allocation between AFS and DYC.

6.2.3.1 DYC Implementation

The additional yaw moment performed by DYC is generated by the differential braking of the vehicle's four wheels. We can find out from Fig. 6.2 that the additional yaw moment from DYC can be mathematically expressed as Eq. (6.34)

$$\Delta M_{DYC} = \sum_{i=1}^{4} (-1)^i F_{xi} L_i \tag{6.34}$$

To achieve DYC's functionality, we must select proper wheels to perform differential braking actions. This section introduces a selection strategy based

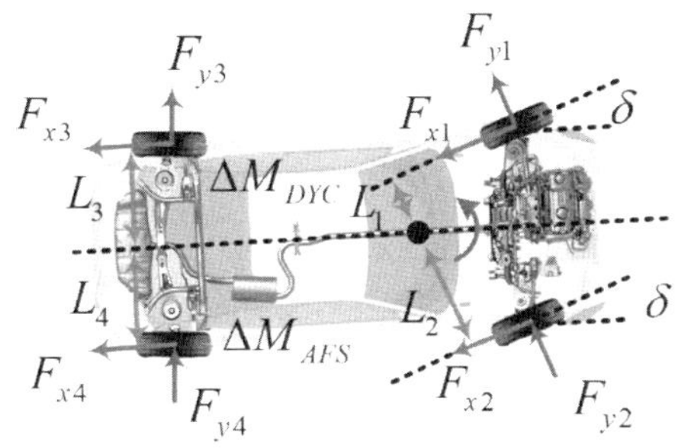

Figure 6.2 Demonstration of the implementation principle of AFS and DYC.

on the relationship between vehicle yaw rate and the desired yaw moment. First, the primary wheel is required to be identified based on ΔM_{DYC} and vehicle dynamic states. If the primary wheel is unable to fulfill the desired yaw moment assigned to DYC, the selection strategy will employ an auxiliary wheel to supplement the required yaw moment.

Concretely, the selection strategy generates a vector representing the active braking status of each wheel based on λ and λ_M, which is defined in Eq. (6.35). The specific strategy is introduced as follows.

1. $\lambda < 0$: The vehicle operates in the condition of oversteering, and the front outer wheel is selected as the primary control wheel.
2. $\lambda > 0$: This suggests that the vehicle is in understeering condition. The rear inner wheel is selected as the primary control wheel.
3. Auxiliary wheel compensation: Regardless of whether the vehicle is operating under oversteering or understeering conditions, an auxiliary wheel is engaged to compensate for the desired yaw moment, when the selected primary control wheel reaches its operational limits.

$$\lambda = \lambda_M \cdot \lambda_\gamma, (\lambda_M = sign(\Delta M_{DYC}), \lambda_\gamma = sign(\gamma)), \tag{6.35}$$

where $sign()$ denotes the signum function, which outputs the sign of a number.

The calculation of the braking pressure for each wheel $P_{B,i}$ is expressed as follows.

$$P_{B,i} = \frac{r_{w,i}}{K_{B,i}} F_{x,i}, \tag{6.36}$$

where $K_{B,i}$ represents the pressure conversion constant of the respective wheel cylinder, and $r_{w,i}$ represents the effective wheel radius.

6.2.3.2 AFS Implementation

First, the lateral tire forces F_{yi} are proportional to their corresponding sideslip angles α_i in a linear tire force model, which is given by

$$F_{yi} = k_i \alpha_i, i = 1, 2, 3, 4. \tag{6.37}$$

where k_i denotes tire cornering stiffness, and tire sideslip angles α_i can be derived as shown in Eq. (6.38) under the assumption that tire sideslip angles remain relatively small.

$$\begin{cases} \alpha_1 = \alpha_2 = \arctan(\beta + \frac{l_f \gamma}{u}) - \delta \approx \beta + \frac{l_f \gamma}{u} - \delta, \\ \alpha_3 = \alpha_4 = \arctan(\beta - \frac{l_r \gamma}{u}) \approx \beta - \frac{l_r \gamma}{u}. \end{cases} \tag{6.38}$$

The additional yaw moment generated by AFS can be expressed by Eq. (6.39).

$$\Delta M_{AFS} = l_f(F_{y1} + F_{y2})\cos\delta - l_r(F_{y3} + F_{y4}) \tag{6.39}$$

Thus, based on the above discussion, the control effort performed by AFS can be reformulated as:

$$\Delta M_{AFS} = (k_f l_f - k_r l_r)\beta + (k_f l_f^2 + k_r l_r^2)\frac{\gamma}{u} - k_f l_f \delta_{AFS} \tag{6.40}$$

6.2.4 EXPERIMENTAL RESULTS AND ANALYSIS

To demonstrate the effectiveness and superior performance of the proposed method, this subsection presents the results and analysis of typical experiments. Besides, several benchmarks including a traditional model predictive control (MPC)-based integrated controller, a traditional DYC controller, and the same experiment without any controller are employed to compare with the proposed method. The test scenario is the double-lane-change (DLC) maneuvers with time-varying vehicle speed. During the DLC procedure, the vehicle speed fluctuates between 85 km/h and 95 km/h, however, it is worth noting that the vehicle loses its dynamic stability and its speed exhibits a divergence at about 4s without any dynamic controller. The comparative analysis highlights the ability of the proposed controller to handle system uncertainties and achieve superior performance in safety-critical conditions.

Vehicle sideslip angle is an essential variable to indicate vehicle lateral stability, and large sideslip angle values usually suggest that the vehicle is in an unstable status. Vehicle sideslip angles controlled by different methods are shown in Fig. 6.3, which illustrates that the proposed coordinated control method outperforms the other controllers. Specifically, the proposed controller can achieve smaller sideslip angles than the other control methods. The maximum sideslip angle achieved by the proposed controller is 2.88°, while the maximum sideslip angles achieved by DYC and the MPC-based coordinated controller are 4.53° and 3.64°, respectively. These results demonstrate the effectiveness of the proposed controller in minimizing sideslip angles and enhancing dynamic stability.

Fig. 6.4 demonstrates vehicle yaw rate controlled by different methods. The results show that the proposed controller can track the reference values of yaw

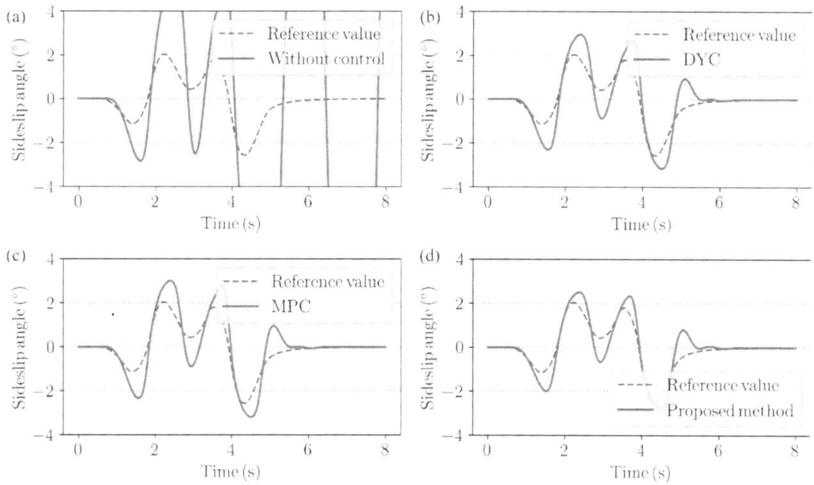

Figure 6.3 Vehicle sideslip angle, (a) the experiment without any control, (b) controlled by DYC, (c) controlled by traditional MPC, (d) controlled by the proposed method.

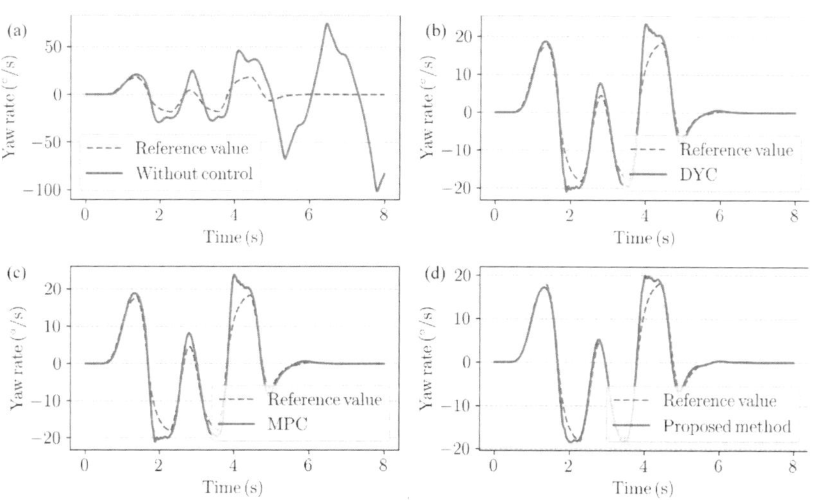

Figure 6.4 Vehicle yaw rate, (a) the experiment without any control, (b) controlled by DYC, (c) controlled by traditional MPC, (d) controlled by the proposed method.

rate more accurately than the other controllers. The vehicle without any controller exhibits a divergent yaw rate curve, which suggests that the vehicle loses its dynamic stability. Even though the tracking errors achieved by DYC and the MPC-based coordinated controller are comparably larger, both controllers can still keep the vehicle stable during the DLC maneuver. In terms of vehicle lateral acceleration, the proposed controller can achieve a smaller lateral acceleration than the other controllers. DYC performs the largest lateral acceleration, which is approximately 7.65 m/s^2, and the maximum lateral acceleration value achieved by the MPC-based coordinated controller is about 7.55 m/s^2. However, the proposed controller achieves the peak lateral acceleration of 7.19 m/s^2. This result also highlights the superior performance of the proposed controller.

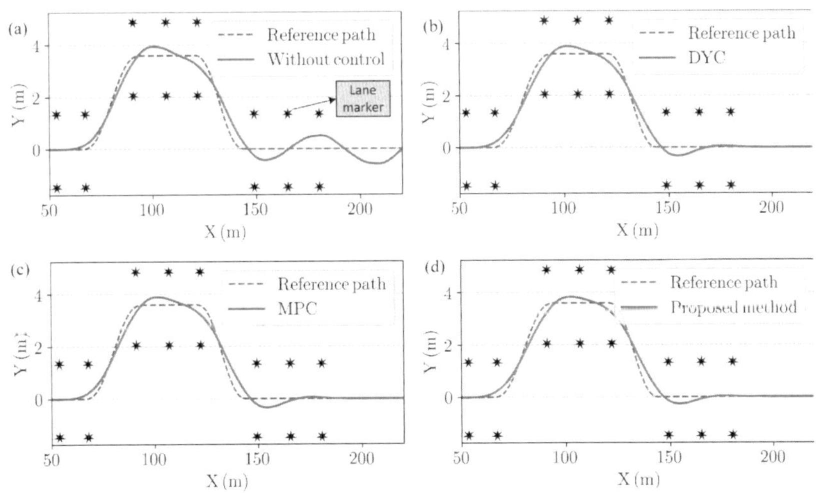

Figure 6.5 Vehicle sideslip angle, (a) the experiment without any control, (b) controlled by DYC, (c) controlled by traditional MPC, (d) controlled by the proposed method.

As demonstrated in Fig. 6.5, the vehicle trajectory could not track the desired path and diverges significantly in the test case without any controller. In contrast, the trajectories performed by the proposed controller, DYC, and the MPC-based coordinated controller can track the desired path accurately. Among these methods, the proposed controller achieves the highest tracking accuracy, which suggests the superior control performance of the proposed method. To sum up, these results illustrate the effectiveness and great performance of the proposed coordinated control method in enhancing vehicle dynamic stability. In addition, both the proposed controller and the MPC-based coordinated controller can perform vehicle dynamic control with lower hydraulic brake pressure than DYC. Specifically, the peak braking pressure performed by DYC reaches approximately 14 MPa, and the MPC-based coordinated controller performs a peak braking pressure of about 13 MPa. In contrast, when the proposed controller is utilized, the peak braking pressure is reduced to about 4 MPa. This reduction in braking efforts can mitigate the adverse effects of braking on riding comfort, which highlights a key advantage of combining AFS into the dynamic control strategy.

6.2.5 SUMMARY

This section presents a robust coordinated control method for integrating AFS and DYC. The proposed method is designed to address the challenge of vehicle parametric uncertainties including vehicle time-varying speed and uncertain mass via a polytopic uncertainty method. LMI technique is employed to derive the coordinated controller. The upper layer generates the desired yaw moment, and then the lower layer distributes the desired yaw moment to AFS and DYC based on vehicle dynamic states. The proposed method is validated by typical experiments, and the results demonstrate that the proposed method can achieve superior performance in guaranteeing vehicle dynamic stability than other methods.

6.3 FRAMEWORK OF CHASSIS DOMAIN APPROACH

As discussed earlier, most of the existing studies on vehicle dynamic control mainly focus on the enhancement of vehicle handling stability. Unlike traditional vehicles which are operated by human drivers, AVs necessitate chassis systems tracking desired trajectories, guaranteeing handling stability, and optimizing the utility of each wheel in a practical way, which presents a significant challenge. Inspired by the potential of fully controllable chassis actuators, which are efficient, reliable, and adaptable [93], this section introduces a global chassis multi-actuator integrated control framework for AVs, named by the chassis domain controller (CDC). Each wheel module is designed to be fully controllable, which can perform driving, braking, and steering actions [142]. Since game theory has been widely applied in system modeling and optimization, the proposed CDC utilizes the game-theoretical paradigm to solve the multi-objective and multi-actuator control optimization problem. Fig. 6.6 shows the diagram of the proposed CDC, and the proposed method sets the foundation for next-generation chassis systems capable of meeting the high-standard demands of autonomous driving.

The mechanical structure of the wheel module is demonstrated in Fig. 6.7. The supporting arm performs the steering torque transmission as well as the dynamic and static load bearing, and the drive system is mounted on the supporting arm, which forms a wheel-side drive configuration. Besides, the steering, suspension, and drive actuators are connected by the supporting arm. The designed wheel module can reduce the unsprung mass and alleviate the heat dissipation problem, so that the vehicle reliability and handling stability can be improved. The wheel module adopts the electromechanical braking system, which is a brake-by-wire technology, to achieve rapid and precise control of braking force.

The operation modes of the wheel-module-based chassis are illustrated in Fig. 6.8. A variety of advanced operations modes can be achieved through different execution combinations of the wheel modules based on driving conditions. For example, the front and rear reverse steering mode can effectively

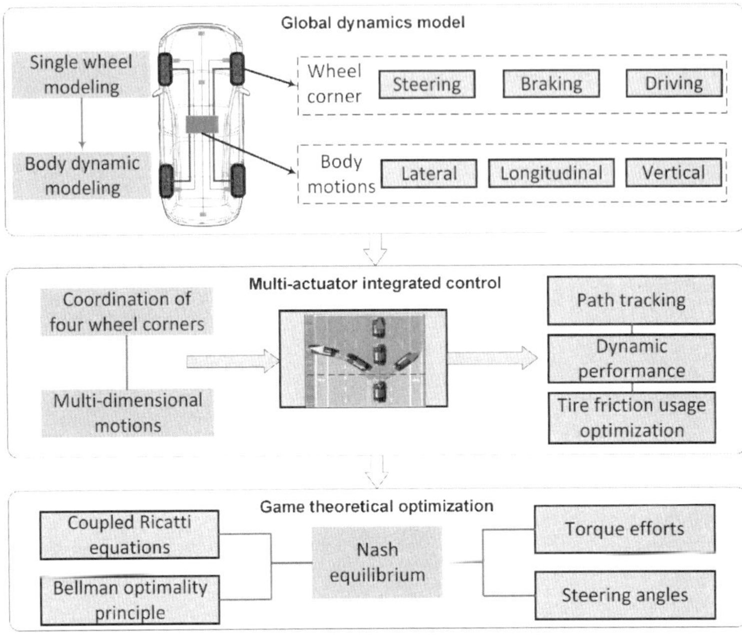

Figure 6.6 The whole architecture of the proposed CDC.

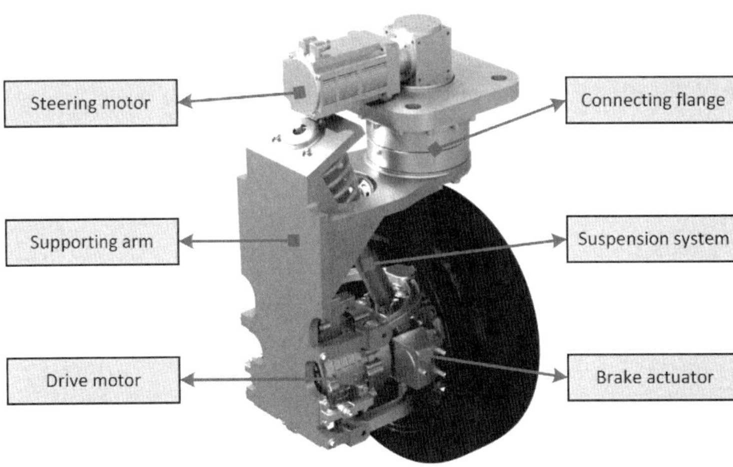

Figure 6.7 Mechanical structure of the wheel module.

reduce the turning radius when a car runs at low speeds, and the oblique driving mode can perform parallel parking or navigating tight spaces via enabling a car to move diagonally. In addition, the zero radius steering and turning around without steering can rotate or change driving direction within its own

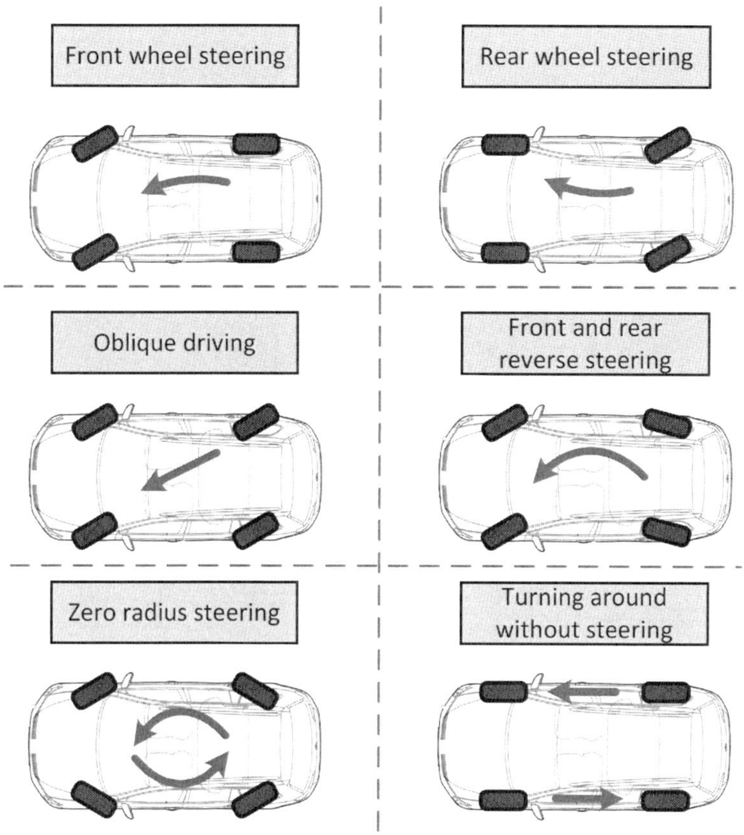

Figure 6.8 Operation modes of the wheel-module-based chassis.

footprint, which provides a highly flexible and improve vehicle's maneuverability.

6.4 VEHICLE MODULAR MODELING

This section first discusses the modeling of the wheel corner, focusing on the single-wheel dynamics and the tire forces acting on the wheel corner. Then, the global vehicle dynamics modeling is introduced, which is based on the assembly of the wheel corner model, as presented in Fig. 6.9.

6.4.1 WHEEL CORNER MODELING

In this section, each wheel corner is treated as an independent and fully controllable unit, which is capable of performing driving, braking, and steering actions. By properly coordinating these actions of all wheel corners, the vehicle

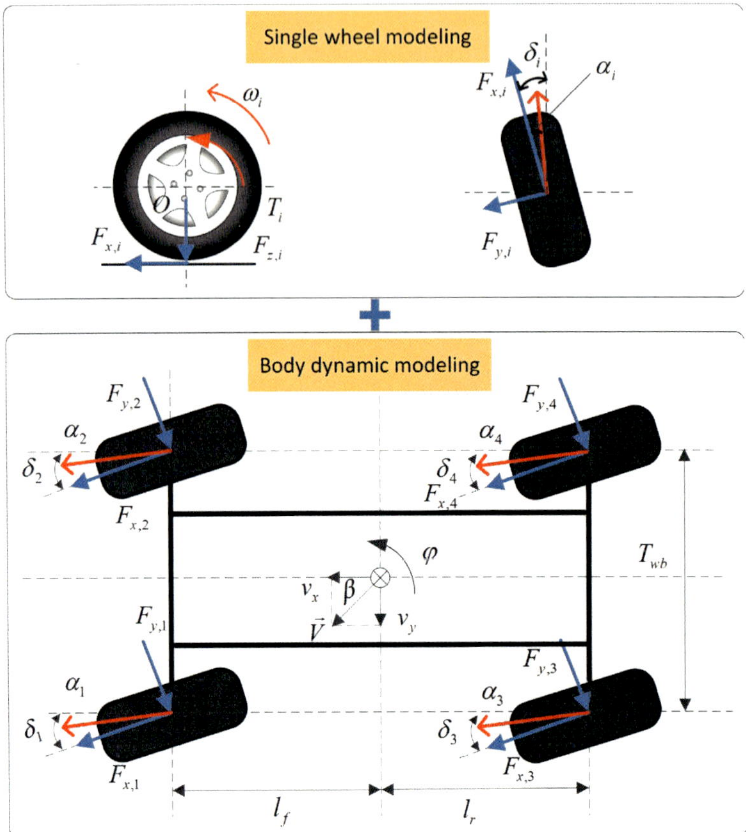

Figure 6.9 Diagram of vehicle global dynamics modeling.

can achieve improved flexibility and maneuverability. The dynamic responses of a vehicle are fundamentally determined by the tire forces at the contact patches. Thus, this section first focuses on the modeling of individual wheel corners. As demonstrated in Fig. 6.9, to ease the representation of the vehicle global dynamic model, this section uses $i = 1, 2, 3, 4$ to represent the front left wheel, front right wheel, rear left wheel, and rear right wheel, respectively. For simplification, T_i is introduced as a unified notation to represent either the traction torque T_{wtij} or braking torque T_{wpij}. The control inputs to each wheel include the traction or braking torque T_i and the steering angle δ_i. These inputs directly influence the tire forces of each wheel, that is, longitudinal tire force $F_{x,i}$, lateral tire force $F_{y,i}$, and vertical tire force $F_{z,i}$.

Tire vertical force acting on each wheel determines the wheel's potential grip, which directly influences the capability of the wheel to maintain traction and guarantee handling stability. The calculation of vertical tire forces has been introduced in Section 3.2.

Each wheel corner has tire longitudinal slip and sideslip phenomenon, which is caused by the elastic deformation inherent to rubber materials. This phenomenon is especially pronounced in the scenarios of critical maneuvers or low-adhesion road surfaces, which is essential to understanding tire dynamics. To represent the tire behaviors, this section introduces the definitions of longitudinal slip ratio λ_i and tire sideslip angle of each wheel corner, which are calculated by Eq. (6.41) and Eq. (6.42), respectively. This section utilizes a simplified linear model to calculate the tire forces acting on each wheel corner, which is expressed in Eq. (6.43). The longitudinal tire force is derived under the assumption that it is proportional to the applied torque, and the lateral tire force is linearly dependent on the tire sideslip angle.

$$\lambda_i = \frac{R \cdot \omega_i - v_x}{\max(R \cdot \omega_i, v_x)} \tag{6.41}$$

$$\alpha_i = \delta_i - \frac{v_y + \varphi \cdot l_i}{v_x} \tag{6.42}$$

where l_i is l_f if $i = 1,2$, otherwise l_i denotes $-l_r$.

$$\begin{cases} F_{x,i} = \dfrac{T_i}{R} \\ F_{y,i} = k_i \cdot \alpha_i \end{cases} \tag{6.43}$$

The rotational dynamics of each wheel corner can be calculated based on the applied torque T_i and the longitudinal tire force $F_{x,i}$, which is expressed in Eq. (6.44). The vehicle might lose its dynamic stability if one wheel exceeds its tire grip, which is determined by the tire–road friction coefficient μ. To keep handling stability, it is essential to ensure that each wheel corner operates within its tire grip limit. To quantify the tire grip utilization, this section introduces γ_i to represent the tire grip margin of each wheel corner, which is calculated by Eq. (6.45).

$$\dot{\omega}_i = \frac{1}{J_{w,i}}(T_i - R \cdot F_{x,i}) \tag{6.44}$$

where $J_{w,i}$ represents the wheel's rotational inertia.

$$\gamma_i = \frac{\sqrt{F_{x,i}^2 + F_{y,i}^2}}{\mu \cdot F_{z,i}} \tag{6.45}$$

The tier forces $F_{x,i}$, and $F_{y,i}$ are defined within their corresponding local tire coordinate systems. To facilitate the vehicle body dynamic modeling, $F_{x,i}$, and $F_{y,i}$ must be transformed into the vehicle body-fixed coordinate system. This coordinate system is centered at the vehicle center of gravity (CG) and it is the reference for analyzing vehicle dynamics in this section. $F_{x,i}$, and $F_{y,i}$ can be resolved into their respective components $F_{vx,i}$ and $F_{vy,i}$, which are expressed in Eq. (6.46). In addition, this section defines that $\boldsymbol{F}_{v,i} = [F_{vx,i}, F_{vy,i}]^T$ and $\boldsymbol{F}_i = [F_{x,i}, F_{y,i}]^T$. By aligning the tire forces to the vehicle body-fixed coordinate system, we integrate the individual contributions from all wheel corners into the global vehicle dynamics model.

$$\begin{bmatrix} \boldsymbol{F}_{v,1} \\ \boldsymbol{F}_{v,2} \\ \boldsymbol{F}_{v,3} \\ \boldsymbol{F}_{v,4} \end{bmatrix} = \underbrace{\begin{bmatrix} \boldsymbol{\Lambda}_1 & 0 & 0 & 0 \\ 0 & \boldsymbol{\Lambda}_2 & 0 & 0 \\ 0 & 0 & \boldsymbol{\Lambda}_3 & 0 \\ 0 & 0 & 0 & \boldsymbol{\Lambda}_4 \end{bmatrix}}_{\boldsymbol{\Lambda}} \underbrace{\begin{bmatrix} \boldsymbol{F}_1 \\ \boldsymbol{F}_2 \\ \boldsymbol{F}_3 \\ \boldsymbol{F}_4 \end{bmatrix}}_{\boldsymbol{F}} \tag{6.46}$$

where

$$\boldsymbol{\Lambda}_i = \begin{bmatrix} \cos\delta_i & -\sin\delta_i \\ \sin\delta_i & \cos\delta_i \end{bmatrix} \tag{6.47}$$

According to the transformation of tire forces, the total forces and moments acting on the vehicle in the body-fixed coordinate system can be calculated as

$$\begin{cases} F_X = \sum\limits_{i=1}^{4} F_{vx,i} \\ F_Y = \sum\limits_{i=1}^{4} F_{vy,i} \\ M_Z = \sum\limits_{i=1}^{4} \left[(-1)^i \frac{b}{2} F_{vx,i} + l_i F_{vy,i}\right] \end{cases} \Rightarrow \begin{bmatrix} F_X \\ F_Y \\ M_Z \end{bmatrix} = \underbrace{\begin{bmatrix} \boldsymbol{\Pi}_1 & \boldsymbol{\Pi}_2 & \boldsymbol{\Pi}_3 & \boldsymbol{\Pi}_4 \end{bmatrix}}_{\boldsymbol{\Pi}} \begin{bmatrix} \boldsymbol{F}_{v,1} \\ \boldsymbol{F}_{v,2} \\ \boldsymbol{F}_{v,3} \\ \boldsymbol{F}_{v,4} \end{bmatrix} \tag{6.48}$$

where $\boldsymbol{\Pi}_i$ is expressed as

$$\boldsymbol{\Pi}_i = \begin{bmatrix} 1 & 0 \\ 0 & 1 \\ (-1)^i \frac{b}{2} & l_i \end{bmatrix} \tag{6.49}$$

6.4.2 VEHICLE BODY DYNAMICS MODELING

This subsection introduces vehicle body dynamic modeling based on the total forces acting on the vehicle body. The planar vehicle body dynamics include longitudinal, lateral, and yaw motions, which is critical for designing vehicle stability control systems and trajectory following algorithms.

1. Longitudinal dynamics: It describes the vehicle's longitudinal acceleration behaviors, The longitudinal dynamics can be expressed as

$$m(a_x - \varphi v_y) = F_X \tag{6.50}$$

2. Lateral dynamics: Vehicle steering responses and lateral acceleration are involved in the lateral dynamics, which is given by

$$m(a_y + \varphi v_x) = F_Y \tag{6.51}$$

3. Yaw dynamics: Vehicle yaw rate is governed by the yaw dynamics, which is expressed as

$$I_z \dot{\varphi} = M_Z \tag{6.52}$$

Thus, given that the body dynamic state vector is defined as $\boldsymbol{x}_b = [v_x, v_y, \varphi]^T$, and the state-space form of the total force is defined as $\boldsymbol{F}_{total} = [F_X, F_Y, M_Z]^T$, the state-space representation of Eq. (6.50)-(6.52) can be expressed as

$$\dot{\boldsymbol{x}}_b = \boldsymbol{A}_b \boldsymbol{x}_b + \boldsymbol{B}_b \boldsymbol{F}_{total} \tag{6.53}$$

where the system matrix $\boldsymbol{A}_b$ and input matrix $\boldsymbol{B}_b$ can be written as

$$\boldsymbol{A}_b = \begin{bmatrix} 0 & 0 & v_y \\ 0 & 0 & -v_x \\ 0 & 0 & 0 \end{bmatrix}, \boldsymbol{B}_b = \begin{bmatrix} m^{-1} & 0 & 0 \\ 0 & m^{-1} & 0 \\ 0 & 0 & I_z^{-1} \end{bmatrix} \tag{6.54}$$

6.4.3 CHASSIS GLOBAL DYNAMICS FORMULATION

A comprehensive representation of vehicle chassis global dynamics is a must for the investigation of a unified control framework. In this section, the global dynamic state includes the dynamics of individual wheel corners and the vehicle body dynamics, which is expressed as

$$\boldsymbol{x}_c = [v_x, v_y, \varphi, \omega_1, \omega_2, \omega_3, \omega_4]^T = [\boldsymbol{x}_b^T, \boldsymbol{x}_w^T]^T \tag{6.55}$$

where the vector $[\omega_1, \omega_2, \omega_3, \omega_4]^T$ denotes the rotational speeds of the four wheels and is referred to as $\boldsymbol{x}_w$.

The proposed CDC aims to incorporate the steering angles and torque inputs of all individually controllable wheels to achieve the desired vehicle dynamic performance. To effectively represent these inputs, we define the control input vector as follows.

$$\boldsymbol{u}_c = [T_1, \delta_1, T_2, \delta_2, T_3, \delta_3, T_4, \delta_4]^T \tag{6.56}$$

The modeling of vehicle chassis domain dynamics requires the integration of the wheel corner dynamics and vehicle body dynamics. Based on the introduction of the wheel corner modeling, this subsection derives the following equation to represent the wheel dynamics.

$$\dot{\boldsymbol{x}}_w = \boldsymbol{A}_w \boldsymbol{x}_w + \boldsymbol{B}_w \boldsymbol{u}_c + \boldsymbol{E}_w \tag{6.57}$$

where $\boldsymbol{A}_w = \mathbf{0}_{4\times4}$. $\boldsymbol{B}_w$ and $\boldsymbol{E}_w$ are given by

$$\boldsymbol{B}_w = \begin{bmatrix} J_{w,1}^{-1} & 0 & 0 & 0 & 0 & 0 & 0 & 0 \\ 0 & 0 & J_{w,2}^{-1} & 0 & 0 & 0 & 0 & 0 \\ 0 & 0 & 0 & 0 & J_{w,3}^{-1} & 0 & 0 & 0 \\ 0 & 0 & 0 & 0 & 0 & 0 & J_{w,4}^{-1} & 0 \end{bmatrix} \tag{6.58}$$

$$\boldsymbol{E}_w = \left[\frac{RF_{x,1}}{-J_{w,1}}, \frac{RF_{x,2}}{-J_{w,2}}, \frac{RF_{x,3}}{-J_{w,3}}, \frac{RF_{x,4}}{-J_{w,4}}\right]^T \tag{6.59}$$

Furthermore, to keep consistency and facilitate seamless integration of different modeling components, this subsection substitutes the tire force transformation described in Eq. (6.48) into the formulation of vehicle body dynamics described in Eq. (6.53) to reformulate vehicle body dynamics, which is expressed as

$$\dot{\boldsymbol{x}}_b = \boldsymbol{A}_b \boldsymbol{x}_b + \boldsymbol{B}_b \boldsymbol{\Pi} \boldsymbol{\Lambda} \boldsymbol{F} \tag{6.60}$$

Both longitudinal and lateral tire forces for all wheel corners, which are expressed within their respective local tire coordinate systems, are included in the vector $\boldsymbol{F}$. It can be derived by integrating the calculation of tire sideslip angle and Eq. (6.43), which is given by

$$\underbrace{\begin{bmatrix} \boldsymbol{F}_1 \\ \boldsymbol{F}_2 \\ \boldsymbol{F}_3 \\ \boldsymbol{F}_4 \end{bmatrix}}_{\boldsymbol{F}} = \underbrace{\begin{bmatrix} \boldsymbol{A}_{t,1} \\ \boldsymbol{A}_{t,2} \\ \boldsymbol{A}_{t,3} \\ \boldsymbol{A}_{t,4} \end{bmatrix}}_{\boldsymbol{A}_t} \boldsymbol{x}_b + \underbrace{\begin{bmatrix} \boldsymbol{B}_{t,1} & \mathbf{0} & \mathbf{0} & \mathbf{0} \\ \mathbf{0} & \boldsymbol{B}_{t,2} & \mathbf{0} & \mathbf{0} \\ \mathbf{0} & \mathbf{0} & \boldsymbol{B}_{t,3} & \mathbf{0} \\ \mathbf{0} & \mathbf{0} & \mathbf{0} & \boldsymbol{B}_{t,4} \end{bmatrix}}_{\boldsymbol{B}_t} \boldsymbol{u}_c \tag{6.61}$$

with

$$\boldsymbol{A}_{t,i} = \begin{bmatrix} 0 & 0 & 0 \\ 0 & \dfrac{k_i}{-v_x} & \dfrac{k_i \cdot l_i}{-v_x} \end{bmatrix}, \boldsymbol{B}_{t,i} = \begin{bmatrix} R^{-1} & 0 \\ 0 & k_i \end{bmatrix} \tag{6.62}$$

Then, this section substitutes the formulation of $\boldsymbol{F}$ into Eq. (6.60) to derive the vehicle body dynamics formulation utilizing $\boldsymbol{u}_c$, which is given by Eq. (6.63). Thus, the comprehensive chassis domain dynamics is established by integrating the body dynamics formulation and the wheel dynamics formulation, which is expressed in Eq. (6.64).

$$\begin{aligned} \dot{\boldsymbol{x}}_b &= \boldsymbol{A}_b \boldsymbol{x}_b + \boldsymbol{B}_b \boldsymbol{\Pi} \boldsymbol{\Lambda} (\boldsymbol{A}_t \boldsymbol{x}_b + \boldsymbol{B}_t \boldsymbol{u}_c) \Rightarrow \\ \dot{\boldsymbol{x}}_b &= (\boldsymbol{A}_b + \boldsymbol{B}_b \boldsymbol{\Pi} \boldsymbol{\Lambda} \boldsymbol{A}_t) \boldsymbol{x}_b + \boldsymbol{B}_b \boldsymbol{\Pi} \boldsymbol{\Lambda} \boldsymbol{B}_t \boldsymbol{u}_c \end{aligned} \tag{6.63}$$

$$\dot{\boldsymbol{x}}_c = \underbrace{\begin{bmatrix} \boldsymbol{A}_b + \boldsymbol{B}_b \boldsymbol{\Pi} \boldsymbol{\Lambda} \boldsymbol{A}_t & \mathbf{0} \\ \mathbf{0} & \boldsymbol{A}_w \end{bmatrix}}_{\boldsymbol{A}_c} \boldsymbol{x}_c + \underbrace{\begin{bmatrix} \boldsymbol{B}_b \boldsymbol{\Pi} \boldsymbol{\Lambda} \boldsymbol{B}_t \\ \boldsymbol{B}_w \end{bmatrix}}_{\boldsymbol{B}_c} \boldsymbol{u}_c + \boldsymbol{E}_c \tag{6.64}$$

where $\boldsymbol{E}_c = [\mathbf{0}_{3\times1}^T, \boldsymbol{E}_w^T]^T$.

6.5 MULTI-ACTUATOR AND MULTI-OBJECTIVE CONTROL

This section introduces the design of the proposed CDC method based on the modeling of chassis domain dynamics. Tire forces of four wheel corners ultimately govern vehicle dynamic behaviors, and the control inputs to each wheel corner are the traction or braking torque and the steering angle. On one hand, vehicle chassis execution is required to fulfill multiple objectives in a variety of driving scenarios. On the other hand, the coordination of these individually controllable wheel corners is challenging, even though the fully controllable wheel corner has substantial advantages in improving vehicle flexibility and maneuverability. Therefore, this section proposes a game-theoretical control approach to address these challenges.

6.5.1 MULTIPLE CONTROL OBJECTIVES

The proposed CDC method is designed to perform accurate trajectory tracking and speed control while maintaining vehicle handling stability. In this section, longitudinal velocity control, path following, dynamic stability performance, and efficient tire force utilization are involved in the control objectives of the proposed CDC method.

6.5.1.1 Handling Stability

Vehicle handling stability can be evaluated by two critical dynamic states, that is, vehicle yaw rate and sideslip angle. The two states must remain within reasonable ranges to ensure safe and stable maneuvers when a vehicle runs in safety-critical conditions. When a vehicle performs steering maneuvers, small sideslip angles and steady yaw rates consistent with the steering angle imply that the vehicle is in a stable status. Furthermore, vehicle sideslip angle β is expressed as $\beta = \arctan(v_y/v_x)$, so this subsection uses v_y in the definition of the handling stability performance function for simplicity and computational efficiency. Thus, to optimize vehicle handling stability, a quadratic performance function Ξ_d is defined based on φ and v_y. The weighting matrix in Ξ_d, denoted as $\boldsymbol{Q}_d$, reflects the relative importance of the two dynamic states in this objective.

6.5.1.2 Trajectory Tracking

This section assumes that the desired velocities v_d, waypoints of desired paths (x_d, y_d), and corresponding curvatures κ_d and heading angles ϕ_d of the desired trajectories are available. The performance of path tracking can be assessed by the deviations at the preview point between the expected path and the actual positions. This section defines that $\tilde{v}$, $\tilde{y}$, and $\tilde{\phi}$ are the speed tracking error, lateral position error, and heading angle error, respectively. These tracking errors are required to be augmented in the system modeling to perform the path tracking task, so this section defines the temporal derivatives of these tracking errors

$$\dot{\tilde{y}} = v_x \cdot \tilde{\phi} - v_y - l_p \cdot \varphi \tag{6.65}$$

$$\dot{\tilde{\phi}} = \kappa_d \cdot v_x - \varphi \tag{6.66}$$

$$\dot{\tilde{v}} = \dot{v}_d - \dot{v}_x \tag{6.67}$$

where l_p represents the preview distance.

Based on the above definitions, the state vector of path tracking can be denoted by $\boldsymbol{x}_p = [\tilde{v}, \tilde{y}, \tilde{\phi}]$. Then, Eqs. (6.65), (6.66), and (6.67) can be reformulated into a matrix form for $\boldsymbol{x}_p$ similar to Eq. (6.64). The path-tracking task is required to minimize the deviations, so this section defines a performance indicator, which is formulated as a quadratic function of $\boldsymbol{x}_p$. This section uses Ξ_p to represent this function, which is given by

$$\Xi_p = \boldsymbol{x}_p^T \cdot \boldsymbol{Q}_p \cdot \boldsymbol{x}_p, \boldsymbol{Q}_p \geq 0 \tag{6.68}$$

where $\boldsymbol{Q}_p$ is the weight matrix of the trajectory tracking state.

6.5.1.3 Tire Usage Optimization

Fundamentally, all wheel corners collectively govern the vehicle's motion via generating specific combinations of longitudinal and lateral tire forces. To perform stable maneuvers, all wheel corners are required to remain within their road adhesion limits. This section introduces an indicator that quantifies the utilization of road adhesion according to the analysis of wheel corner dynamics as well as Eq. (6.45), and its expression is given by Eq. (6.69). When a car operates on roads with specific road friction coefficients, lower utilization of road adhesion corresponds to better vehicle handling stability performance. Thus, this section achieves tire usage optimization by minimizing Eq. (6.69). To perform this optimization, we define the performance indicator of tire usage optimization for each wheel corner based on Eq. (6.69), as shown in Eq. (6.70).

$$\gamma_i^2 = \frac{F_{x,i}^2 + F_{y,i}^2}{(\mu \cdot F_{z,i})^2} \Rightarrow \frac{1}{(\mu \cdot F_{z,i} \cdot R)^2} T_i^2 + \left(\frac{k_i}{\mu \cdot F_{z,i}}\right)^2 \alpha_i^2 \tag{6.69}$$

$$\Xi_{t,i} = [T_i, \delta_i] \cdot \boldsymbol{R}_{t,i} \cdot [T_i, \delta_i]^T \tag{6.70}$$

where $\boldsymbol{R}_{t,i}$ represents the weight matrix corresponding to the input at each wheel corner.

6.5.2 CONTROLLER DESIGN

Game theory has been widely applied to strategic behavior modeling, and the decisions of all participating players affect the outcome of all players.

Fundamentally, a dynamic game consists of three components:

1. All involved players in the game.
2. The cost functions associated with each player's objectives.
3. The control strategies adopted by each player.

In this section, each wheel corner can perform braking, driving, and steering. However, certain maneuvers exert a considerable on specific control objectives but have comparatively minor effects on other objectives. Thus, this section defines M players, which consist of corresponding control inputs, according to the above analysis of control objectives. The objective functions are then reformulated and allocated to their corresponding players. In addition, the trajectory tracking formulations as shown in Eqs. (6.67)–(6.66) are augmented to the chassis domain dynamic modeling as shown in Eq. (6.64), and then the following formula can be derived

$$\begin{aligned}&\underbrace{\begin{bmatrix}\dot{\boldsymbol{x}}_p\\ \dot{\boldsymbol{x}}_c\end{bmatrix}}_{\dot{\boldsymbol{x}}_g}=\underbrace{\begin{bmatrix}\boldsymbol{A}_p & \\ \boldsymbol{0}_{7\times3} & \boldsymbol{A}_c\end{bmatrix}}_{\boldsymbol{A}_g}\underbrace{\begin{bmatrix}\boldsymbol{x}_p\\ \boldsymbol{x}_c\end{bmatrix}}_{\boldsymbol{x}_g}+\underbrace{\begin{bmatrix}\boldsymbol{B}_p\\ \boldsymbol{B}_c\end{bmatrix}}_{\boldsymbol{B}_g}\boldsymbol{u}_c+\underbrace{\begin{bmatrix}\boldsymbol{0}\\ \boldsymbol{E}_c\end{bmatrix}}_{\boldsymbol{E}_g}\\ &\Rightarrow \dot{\boldsymbol{x}}_g=\boldsymbol{A}_g\cdot\boldsymbol{x}_g+\boldsymbol{B}_g\cdot\boldsymbol{u}_g+\boldsymbol{E}_g\end{aligned} \tag{6.71}$$

where $\boldsymbol{A}_p$ can be yielded based on Eqs. (6.65)–(6.67).

In this chassis domain control approach, this chapter defines three players based on the chassis actuators and chassis dynamic control objectives:

1. $\boldsymbol{u}_1$: this player involves the steering angles of the front wheel corners and its utilization function performs the optimization of lateral path tracking.
2. $\boldsymbol{u}_2$: this player involves the steering angles of the rear wheel corners and its utilization function performs the optimization of chassis dynamic stability.
3. $\boldsymbol{u}_3$: this player involves the torques of four wheel corners and its utilization function performs the optimization of longitudinal speed tracking.

Moreover, tire usage optimization is also integrated into the control design, and it is combined with the performance indicators discussed in the former subsection. Thus, this section defines the utilization functions Υ_1, Υ_2, and Υ_3 for the corresponding player $\boldsymbol{u}_1$, $\boldsymbol{u}_2$, and $\boldsymbol{u}_3$, respectively.

Based on the above definition, this section defines the corresponding control matrices $\boldsymbol{B}_i$ for the control input of each player $\boldsymbol{u}_i$ through extracting relevant components from $\boldsymbol{B}_g$. Thus, the overall control term $\boldsymbol{B}_g\cdot\boldsymbol{u}_g$ can be calculated by the sum of actions of all participating players, which is expressed as $\boldsymbol{B}_g\cdot\boldsymbol{u}_g=\sum_{i=1}^{3}\boldsymbol{B}_i\cdot\boldsymbol{u}_i$.

Then, the utilization functions of all players are formulated as infinite-horizon optimization functions, which are expressed as

$$\begin{aligned} &\Upsilon_1(\boldsymbol{x}_g, \boldsymbol{u}_1, \boldsymbol{u}_2, \boldsymbol{u}_3) = \int_0^\infty \left(\Xi_{p|\tilde{v}} + \sum_{i=1}^{2} \Xi_{t,i|\delta} \right) dt \\ &\Rightarrow \Upsilon_1 = \int_0^\infty \left(\boldsymbol{x}_g^T \boldsymbol{Q}_1 \boldsymbol{x}_g + \sum_{j=1}^{3} \boldsymbol{u}_j^T \boldsymbol{R}_{1j} \boldsymbol{u}_j \right) dt \end{aligned} \tag{6.72}$$

$$\begin{aligned} &\Upsilon_2(\boldsymbol{x}_g, \boldsymbol{u}_1, \boldsymbol{u}_2, \boldsymbol{u}_3) = \int_0^\infty \left(\Xi_{p|\tilde{y},\tilde{\phi}} + \sum_{i=3}^{4} \Xi_{t,i|\delta} \right) dt \\ &\Rightarrow \Upsilon_2 = \int_0^\infty \left(\boldsymbol{x}_g^T \boldsymbol{Q}_2 \boldsymbol{x}_g + \sum_{j=1}^{3} \boldsymbol{u}_j^T \boldsymbol{R}_{2j} \boldsymbol{u}_j \right) dt \end{aligned} \tag{6.73}$$

$$\begin{aligned} &\Upsilon_3(\boldsymbol{x}_g, \boldsymbol{u}_1, \boldsymbol{u}_2, \boldsymbol{u}_3) = \int_0^\infty \left(\Xi_d + \sum_{i=1}^{4} \Xi_{t,i|T} \right) dt \\ &\Rightarrow \Upsilon_3 = \int_0^\infty \left(\boldsymbol{x}_g^T \boldsymbol{Q}_3 \boldsymbol{x}_g + \sum_{j=1}^{3} \boldsymbol{u}_j^T \boldsymbol{R}_{3j} \boldsymbol{u}_j \right) dt \end{aligned} \tag{6.74}$$

where the corresponding weighting matrix for each quadratic term can be derived based on the definitions of these performance indicators. Specifically, Q_i is defined as the weighting matrix corresponding to the first quadratic term, while R_{ij} represents the weighting matrix corresponding to the second quadratic term. These matrices Q_i and R_{ij} are explicitly assigned to align with Υ_i for each player.

First, this section introduces the definition of Nash equilibrium as follows.

Definition 6.3 *A set of control inputs, which is denoted as* $(\boldsymbol{u}_1^{Nash}, \boldsymbol{u}_2^{Nash}, ..., \boldsymbol{u}_M^{Nash})$, *is defined as a Nash equilibrium solution in the context of an M-player game if, for every player i,* $1 \le i \le M$, *the condition presented in Eq.* (6.75) *is satisfied for any alternative control inputs* $\boldsymbol{u}_i$. *The selected Nash solution* $\boldsymbol{u}_i^{Nash}$ *minimizes the utilization function of player i, assuming that all other players* $j \neq i$ *adhere to their corresponding Nash control inputs* $\boldsymbol{u}_j^{Nash}$. *This suggests that no single player can decrease their utilization function by deviating from their Nash control input, thereby constructing a status of strategic balance among all players.*

$$\Upsilon_i(\boldsymbol{x}_g, \boldsymbol{u}_1^{Nash}, ..., \boldsymbol{u}_i^{Nash}, ..., \boldsymbol{u}_M^{Nash}) \le \Upsilon_i(\boldsymbol{x}_g, \boldsymbol{u}_1^{Nash}, ..., \boldsymbol{u}_i, ..., \boldsymbol{u}_M^{Nash}) \tag{6.75}$$

The control input $\boldsymbol{u}_i$ of each player performs actions independently while being affected by the actions of other players, which is represented by the concept of Nash equilibrium. According to the Hamiltonian definition and Bellman's optimality principle [132], this section derives the Nash optimal control strategies for all players as follows.

$$\boldsymbol{u}_i^{Nash} = -\frac{1}{2} \boldsymbol{R}_i^{-1} \boldsymbol{B}_i^T \boldsymbol{P}_i \boldsymbol{x}_g(t) \tag{6.76}$$

where $\boldsymbol{P}_i$ represents a positive definite matrix, which is determined by solving a set of coupled algebraic matrix Riccati equations, which is given by Eq. (6.77). This numerical solution process can achieve the Nash equilibrium through the interdependence of the involved control strategies.

$$0 = \boldsymbol{P}_i \boldsymbol{A}_c + \boldsymbol{A}_c^T \boldsymbol{P}_i + \boldsymbol{Q}_i + \sum_{j=1}^{3} \boldsymbol{P}_j \boldsymbol{B}_j \boldsymbol{R}_{jj}^{-T} \boldsymbol{R}_{ij} \boldsymbol{R}_{jj}^{-1} \boldsymbol{B}_j^T \boldsymbol{P}_j \tag{6.77}$$

where

$$\boldsymbol{A}_c = \boldsymbol{A}_g - \sum_{i=1}^{3} \boldsymbol{B}_i \boldsymbol{R}_{ii}^{-1} \boldsymbol{B}_i^T \boldsymbol{P}_i \tag{6.78}$$

The chassis actuators can be effectively coordinated to achieve optimal performance of multiple chassis objectives via the Nash control inputs $\boldsymbol{u}_i^{Nash}$, which are derived based on the solution of the coupled algebraic matrix Riccati equations. The Nash equilibrium ensures that each player's control strategy is optimal given the strategies of the other players, leading to a balanced and efficient control of chassis domain dynamics.

6.6 TESTING RESULTS AND ANALYSIS

To demonstrate the effectiveness of the proposed control approach, this section conducts a series of experimental tests, for example, the double lane change (DLC) and the cruising scenario on curvy roads. The performance of the proposed method is compared against two commonly deployed methods: the model-based compensation algorithm (MBCA) and the linear quadratic regulator (LQR) [89]. Both of them have been widely utilized in vehicle four-wheel-driving and front-wheel-steering control. The LQR controller usually minimizes a quadratic cost function subject to system dynamic constraints. Moreover, the MBCA method is designed based on compensatory control and is derived from the system modeling. For the LQR method, its state weighting matrix is defined as diag[200, 200, 0.1, 0.0001, 0.1, 0.1], and its input weighting matrix is diag[0.0001, 80]. Meanwhile, the MBCA method uses compensation factors of 600 for longitudinal control and 0.04 for lateral control. In addition, the key parameters of the proposed method are set as presented in Table. 6.1 in this section.

6.6.1 CRUISING SCENARIO ON CURVY ROADS

This subsection presents the test results of the cruising scenario on curvy roads, where vehicle velocity changes between approximately 130 km/h and 65 km/h. The maximum lateral offset in this scenario is 2.3 m. The trajectory tracking results suggest that the proposed CDC method, LQR method, and MBCA method are capable of following the desired trajectories, which are demonstrated in Fig. 6.10. Moreover, the proposed CDC method outperforms

Table 6.1
Key parameters of the proposed CDC method.

Parameters	Values
Q_1	$\text{diag}[\epsilon_1, \epsilon_0, \epsilon_0, \epsilon_0, \epsilon_0, \epsilon_0]$
Q_2	$\text{diag}[\epsilon_0, \epsilon_2, \epsilon_0, \epsilon_0, \epsilon_0, \epsilon_0]$
Q_3	$\text{diag}[\epsilon_0, \epsilon_0, \epsilon_0, \epsilon_0, \epsilon_3, \epsilon_3]$
R_{11}	$\text{diag}[\rho_1/F_{z,1}, \rho_1/F_{z,2}, \rho_1/F_{z,3}, \rho_1/F_{z,4}]$
R_{22}	$\text{diag}[\rho_2, \rho_2]$
R_{33}	$\text{diag}[\rho_3, \rho_3]$

$\epsilon_0 = 1e^{-5}$, $\epsilon_1 = 300$, $\epsilon_2 = 300$, and $\epsilon_3 = 0.02$.
$\rho_1 = 2$, $\rho_2 = 300$, and $\rho_3 = 200$.

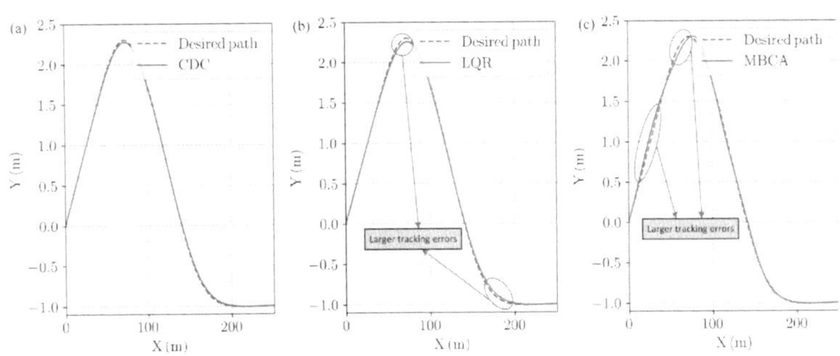

Figure 6.10 Comparison of trajectories in the curvy cruising scenario, (a) the proposed CDC method, (b) the LQR controller, (c) the MBCA method.

the other methods, and the MBCA method achieves the worst path tracking accuracy among the three methods. Therefore, the following discussion about this scenario does not involve the MBCA method. Fig. 6.11 illustrates corresponding vehicle velocities. It is observed that the proposed CDC controller achieves superior performance in tracking the desired vehicle speeds, which is particularly evident during the initial 4s of the cruising scenario.

Both the vehicle yaw rate and sideslip angle exhibit similar trends during the first 10s of the cruising test when the vehicle is controlled by either the proposed CDC method or LQR controller, however, large deviations between the two controllers become evident after 10 s. This divergence indicates a growing tendency toward dynamic instability when the vehicle is controlled by the LQR controller. To analyze this phenomenon further, this subsection will discuss the tire forces and tire grip usage in the cruising test.

Figs. 6.12, 6.13, and 6.14 demonstrate all tire forces performed by the proposed CDC method and LQR controller. Notably, it is observed from Fig. 6.15 that the rear tires achieve significantly high tire grip utilizations during the

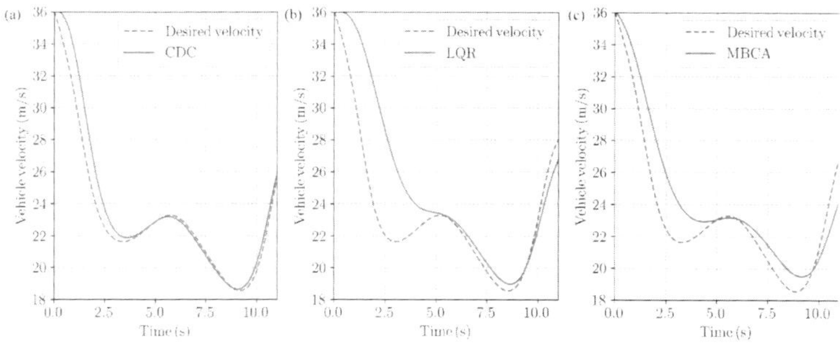

Figure 6.11 Comparison of vehicle velocities in the curvy cruising scenario, (a) the proposed CDC method, (b) the LQR controller, (c) the MBCA method.

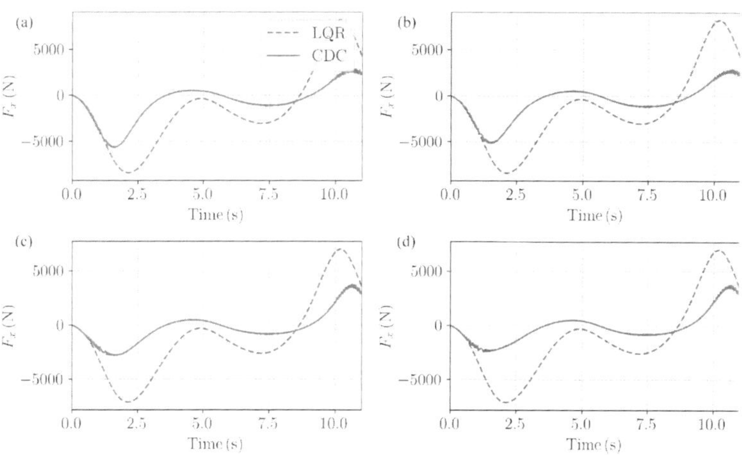

Figure 6.12 Longitudinal tire forces controlled by different methods in the curvy cruising scenario, (a) the left front wheel, (b) the right front wheel, (c) the left rear wheel, (d) the right rear wheel.

initial 5 s. Furthermore, the grip utilizations of the front tires reach an abnormal level between approximately 9 s and 11 s. This phenomenon in grip usage drives the tendency toward dynamic instability after 10s when the vehicle is controlled by the LQR method.

The conventional LQR method lacks the ability to dynamically adjust control commands for individual wheel corners based on their real-time conditions. As a result, when using the LQR method, the front wheel corners exceed their corresponding tire adhesion limits, which causes the loss of steering capability. In contrast, the proposed CDC method can effectively perform the precise tracking of both the vehicle velocities and lateral paths under the premise of optimizing the tire grip usage for each wheel corner, as demonstrated in Fig. 6.15.

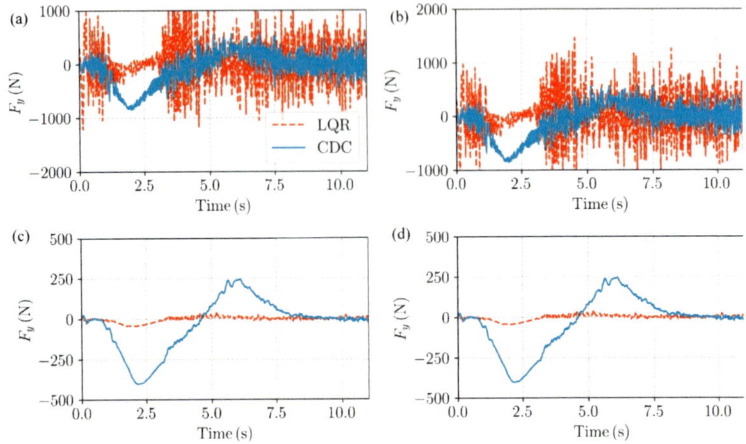

Figure 6.13 Lateral tire forces controlled by different methods in the curvy cruising scenario, (a) the left front wheel, (b) the right front wheel, (c) the left rear wheel, (d) the right rear wheel.

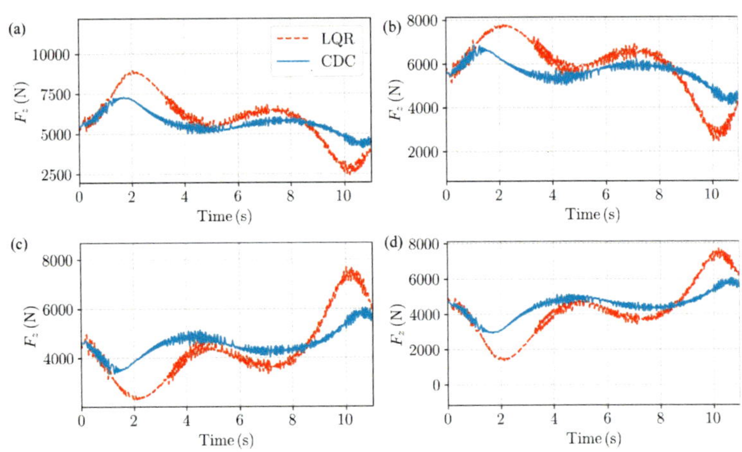

Figure 6.14 Vertical tire forces controlled by different methods in the curvy cruising scenario, (a) the left front wheel, (b) the right front wheel, (c) the left rear wheel, (d) the right rear wheel.

The proposed CDC method ensures that the tire grip usage of each wheel corner remains within its corresponding road adhesion limit. Grip usage of each wheel corner is controlled below 1 with the peak value of approximately 0.75 at the front wheel corner. It is observed that only the front wheel corners are engaged to perform steering operations during the first steering maneuver. Moreover, all wheel corners perform steering operations during the second steering maneuver. Results demonstrate that the proposed CDC method

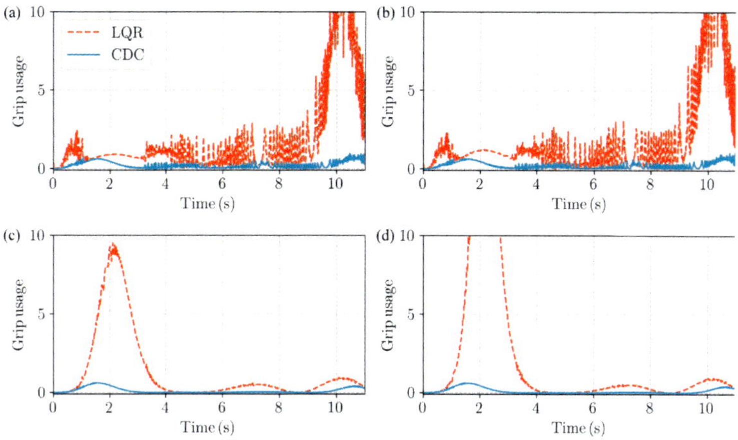

Figure 6.15 Tire grip usage performed by different methods in the curvy cruising scenario, (a) the left front wheel, (b) the right front wheel, (c) the left rear wheel, (d) the right rear wheel.

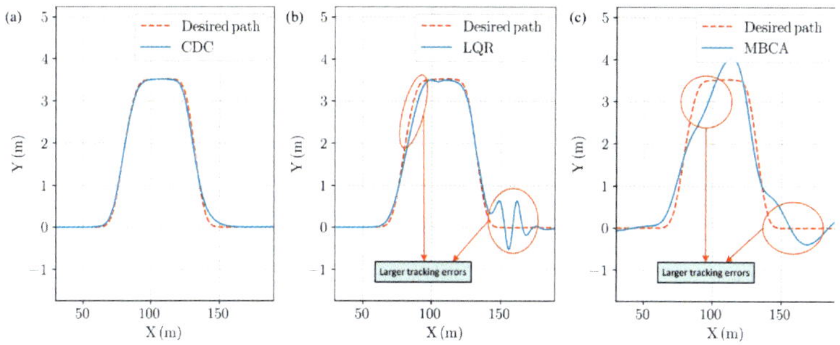

Figure 6.16 Comparison of trajectories in the DLC scenario, (a) the proposed CDC method, (b) the LQR controller, (c) the MBCA method.

outperforms the conventional LQR controller in optimizing chassis domain dynamics and achieving effective multi-actuator coordinated control.

6.6.2 DOUBLE LANE CHANGE SCENARIO

This DLC test is conducted with an initial velocity of 120 km/h, and a sequence of maneuvers is performed, including deceleration, lane-changing, and acceleration, as shown in Figs. 6.16 and 6.17. The results demonstrate that the proposed CDC method outperforms both the LQR controller and the MBCA method in tracking the desired paths and the desired velocities. Furthermore, the MBCA method achieves the worst performance in this DLC test, and this subsection does not discuss the MBCA method in the following analysis.

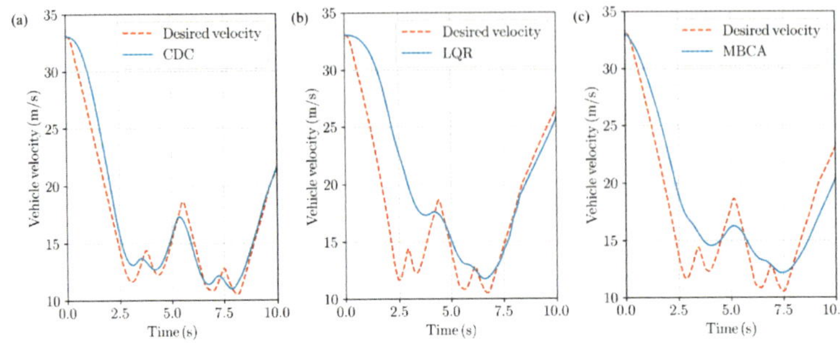

Figure 6.17 Comparison of vehicle velocities in the DLC scenario, (a) the proposed CDC method, (b) the LQR controller, (c) the MBCA method.

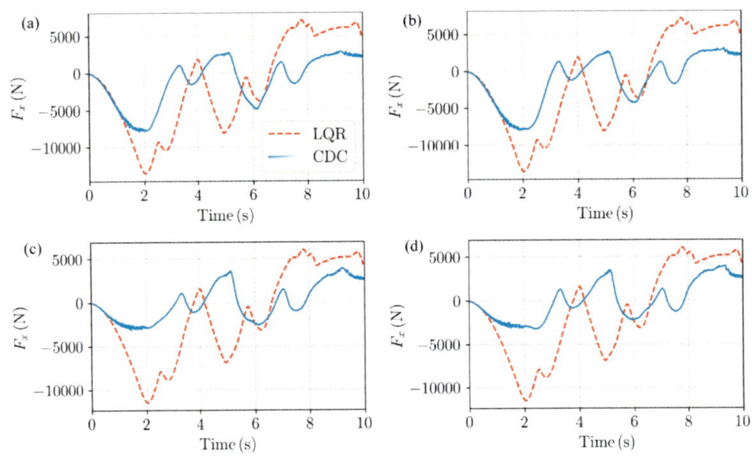

Figure 6.18 Longitudinal tire forces controlled by different methods in the DLC scenario, (a) the left front wheel, (b) the right front wheel, (c) the left rear wheel, (d) the right rear wheel.

Vehicle yaw rate and sideslip angle diverge at about 5 s when the vehicle is controlled by the LQR controller during this test, suggesting that the vehicle loses its dynamic stability. The proposed CDC method performs with reasonable values of vehicle yaw rate and sideslip angle. This illustrates that the proposed method can output suitable control commands to guarantee handling stability. These results highlight the superior performance of the proposed CDC method in path tracking, speed regulation, and optimization of dynamic stability.

The tire forces in this DLC scenario are illustrated in Figs. 6.18, 6.19, and 6.20. As observed in Fig. 6.21, vertical tire forces at each wheel corner experience fluctuations due to load transfer caused by vehicle lateral and longitudinal accelerations. This load variation affects the grip potential of

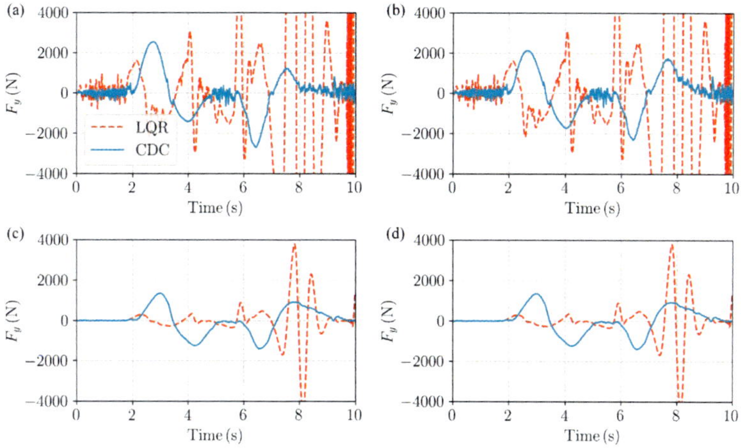

Figure 6.19 Lateral tire forces controlled by different methods in the DLC scenario, (a) the left front wheel, (b) the right front wheel, (c) the left rear wheel, (d) the right rear wheel.

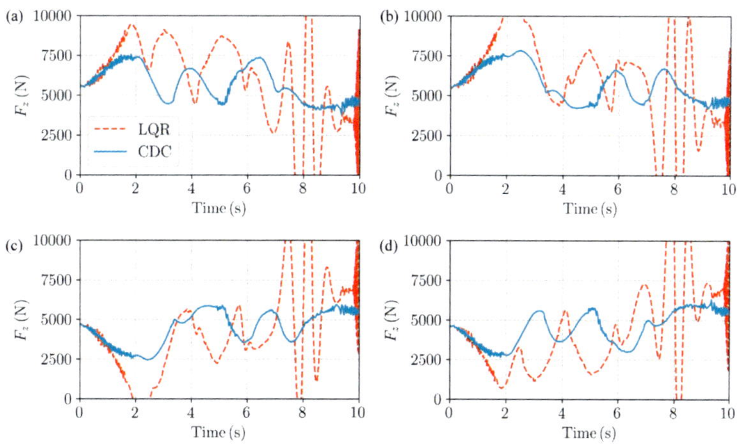

Figure 6.20 Vertical tire forces controlled by different methods in the DLC scenario, (a) the left front wheel, (b) the right front wheel, (c) the left rear wheel, (d) the right rear wheel.

all wheel corners. Fig. 6.21 demonstrates the tire grip usage of each wheel corner. It can be observed that the tire grip usage controlled by the proposed CDC method is consistently balanced across all wheels. Each wheel maintains a reasonable grip usage, with a peak value of below 1. This suggests the proposed method's capability to optimize tire grip when performing multi-actuator control. Conversely, the rear tire grip usage controlled by the LQR method exhibits obvious instability, with sharp fluctuations and peak values

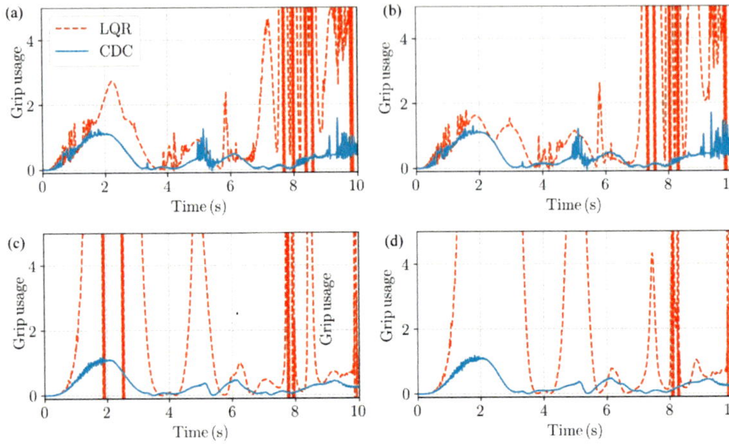

Figure 6.21 Tire grip usage performed by different methods in the DLC scenario, (a) the left front wheel, (b) the right front wheel, (c) the left rear wheel, (d) the right rear wheel.

reaching an unreasonable value of 10. This indicates the infeasibility of the traditional LQR method in chassis-coordinated control.

Since the traditional LQR method performs improper allocation of motor torque efforts, the vehicle fails to track its desired velocities and paths accurately due to tire adhesion saturation. In contrast, the proposed CDC method can effectively perform coordinated control of all wheel corners, significantly improving both path-tracking accuracy and chassis dynamic stability. Specifically, in this DLC scenario, the front wheel corners achieve a peak steering angle of about $-4.8°$ at 6.5 s, while the rear wheel corners reach a maximum angle of approximately $0.12°$ at 6.2 s.

6.6.3 SUMMARY

This section introduces a global chassis integrated control method for AVs. This method incorporates a general global chassis dynamic model based on fully controllable wheel corners. The proposed method is designed to address the issue of chassis multi-objective control including path tracking, speed control, and the optimization of chassis dynamic stability. To achieve this, this section introduces a game-theoretical control strategy to coordinate vehicle braking, driving, and steering. Thus, the proposed method can perform chassis multi-objective, multi-actuator control by employing the Nash equilibrium solution. Extensive experiments are presented to demonstrate the adaptability and effectiveness of the proposed method. The results highlight the superior performance of the proposed method. The proposed method holds substantial promise for advancing the deployment of AVs and merits further exploration.

6.7 CONCLUSION

This chapter introduces two advanced methods for enhancing vehicle dynamic control. First, this chapter adopts a polytopic uncertainty approach to address vehicle parametric uncertainties. AFS and DYC are integrated through the LMI-based controller design to improve vehicle dynamic stability. Second, this chapter introduces a global chassis coordinated control based on a game-theoretical framework to achieve multi-objective control via the coordination of braking, driving, and steering. Typical experimental validations demonstrate the superior performance and adaptability of the proposed methods. These two methods lay a solid foundation for future innovations in the development of AVs.

7 Conclusion and Prospects

Previous chapters have introduced detailed achievements of vehicle dynamics in the context of autonomous driving, including dynamic observation, dynamic modeling, motion planning, and chassis-integrated control. Despite these advancements, the deployment of fully autonomous driving remains a considerable challenge, and substantial work is still required. This chapter will summarize the current progress and provide a brief discussion on prospects, highlighting the ongoing efforts and potential directions for further research and development in the field of autonomous driving.

7.1 CONCLUSIONS

This book extensively addresses the critical aspect of safe and stable operation in AVs, specifically focusing on the chassis implementation for autonomous driving. In-depth research has been conducted on the observation of vehicle dynamic states and the key technologies for controlling chassis dynamic stability. Based on a comprehensive literature review, this book analyzed the coupling relationship between various chassis functions and chassis dynamic stability, as well as the interactions and operational characteristics of chassis subsystems. Following this analysis, a comprehensive framework for chassis dynamic domain control tailored to AVs was proposed. The logical process for chassis domain control was designed, with a thorough investigation into its key technological aspects. The book tackled various issues including the observation of tire dynamic parameters, estimation of key chassis dynamic states, and multi-objective, multi-subsystem coordinated control. Hence, This book aims to overcome the limitations on the estimation accuracy of dynamic states due to the nonlinearity of vehicle and tire dynamics, and the coordination challenges posed by complex and uncertain traffic conditions, thereby enhancing the stability and safety of autonomous driving.

1. To address the common challenges faced by the AV sector, we proposed a comprehensive framework for chassis dynamics domain control tailored to AVs. This framework addresses the problems of model mismatches and control inaccuracies in chassis dynamics. After analyzing the coupling relationships between chassis dynamic stability and other chassis functions, as well as the interactions and operational characteristics of chassis subsystems, we provided the main components of the chassis domain control framework. The logical process of the chassis domain control was also discussed. Then, prevailing vehicle dynamic models including the four-wheel 7DOF dynamic model and single-track 2DOF dynamic model were introduced.

DOI: 10.1201/9781003583929-7

2. Regarding the estimation of tire dynamic parameters, we proposed a method for calculating tire vertical forces based on vehicle longitudinal and lateral accelerations. Subsequently, a unified ST-SMO system for tire dynamic states was designed. Correspondingly, the observers for longitudinal tire forces, lateral tire forces, and tire sideslip angles were designed based on the ST-SMO algorithm. To validate the proposed observation system, we developed a comprehensive dynamic state observation platform. This platform integrates sensors such as WFTs and optical SFII sensors. Experimental results demonstrated that the proposed methods for observing tire dynamic states can accurately estimate tire forces and tire sideslip angles, showing significant advantages over methods like EKF-based estimators.
3. We designed a zero-reset integration method based on sensor signals and an ASCKF estimation method for observing vehicle sideslip angles. These methods are adaptively fused based on vehicle dynamic states to achieve a precise estimation of vehicle sideslip angle through the adaptive fusion algorithm. To address the challenge of estimating road surface adhesion coefficient under complex road conditions, we designed a mode-switching observer for the observation of road friction coefficient. This observer employs the RLS identification method and a compensation algorithm to comprehensively estimate the road adhesion coefficient. These methods significantly improve the accuracy of observing dynamic states including vehicle sideslip angle.
4. The rigorous steady boundaries of vehicle dynamics states were exposed mathematically based on dynamics modeling and vehicle spatio-temporal signal analysis. It provides a benchmark for AVs' inner dynamics safety. The strict mathematical derivation of the evaluation methodology for assessing a vehicle's comprehensive chassis performance was presented, and a rigorous evaluation metric for AVs' domain dynamics was designed. Then, we proposed the chassis-domain-aware trajectory planning method and presented the multi-objective optimization formulation. Geometrical smoothness, chassis global dynamics responses, and feasible safety are all covered in this optimization formulation. The proposed chassis global dynamics-oriented planning scheme exploits a predesigned numerical optimization algorithm to solve this trajectory planning problem. Multiple simulation tests were carried out to evaluate its performance, and experimental results were analyzed and validated the feasibility and efficiency of the proposed scheme.
5. We formulated the global chassis dynamic model by defining each wheel to be fully controllable including braking, driving, and steering. Then, this book developed a game-theoretical control scheme to realize the coordinated control of multiple chassis actuators, and the Nash equilibrium solution was solved to optimize multiple objectives.

Experimental results illustrated the feasibility and effectiveness of our proposed method and its dramatic promotion of AV's deployment.

7.2 PROSPECTS

Autonomous driving technologies represent the forefront of automotive technological innovation. However, the highly complex and dynamic traffic environments, coupled with uncertain and nonlinear dynamic characteristics and the interactions of multiple chassis actuators, pose significant challenges to the safe and stable operation of AVs. This book has conducted a preliminary exploration into the chassis domain control for AVs' chassis dynamics. Further research is still needed to develop the chassis domain controller and its architecture. Future work needs to be conducted to further elucidate the calculation of chassis dynamic states and human-driver-like preview mechanisms. We will further investigate the functional definitions and logical architecture of AVs' decision-making and chassis domain controller to facilitate the real-car deployment of the chassis domain controller. Simultaneously, we will explore the high-computing power, multi-source interface hardware architecture, and design for AV's chassis domain controller. Additionally, we briefly list some other points which can be further investigated from the research in this book.

1. This book primarily focused on the design of observation algorithms for core state information related to chassis stability. Further development should be directed toward the design of observation algorithms for additional vehicle parameters and dynamic state variables such as vehicle mass, pitching rate, rolling rate, and vehicle vertical bounce. This will establish a comprehensive vehicle dynamic acquisition system. An in-depth exploration of the real-time calculation and preview mechanisms of a 15DOF vehicle dynamics model is necessary to further enhance various control performances related to vehicle dynamic stability and comfort. This will also enable the digitalization of AVs' chassis dynamics and achieve self-quantitative assessment of AVs' chassis dynamics.
2. This book proposed an evaluation metric for AVs' domain dynamics performance. Further study could concentrate on its application to real cars to identify different vehicle styles and driving environments. The limitation of our research is the relationship analysis between occupants' subjective feelings and our proposed metric. Therefore, future works can be carried out from the following aspect: recruit participants for scoring vehicle performance subjectively, and analyze the subjective property of our quantitative metric methodology.
3. We have developed the chassis-domain-aware trajectory planning method, and we also proposed a game-theoretical control scheme to realize the coordinated control of multiple chassis actuators. The shortcomings of this work on trajectory planning and integrated

control lie in the absence of real-car tests due to the limitations of experimental cars and conditions. For future work, we will focus on the manufacturing of the fully controllable wheel prototype. Then, we will investigate the real-car application of the proposed methods. Additionally, we will investigate the variations in the application of our proposed framework depending on specific vehicle configurations.

References

1. Ahmad Noor Syukri Zainal Abidin, Arief Hakimi Azmi, Khairil Anwar Abu Kassim, Ahmad Shahir Jamaludin, and Mohd Nizar Mhd Razali. A review on automotive tires significant characteristic identification for general consumers. In *Proceedings of the 2nd Energy Security and Chemical Engineering Congress: Selected Articles from ESChE 2021, Malaysia*, pages 375–385. Springer, 2022.
2. Joel A E Andersson, Joris Gillis, Greg Horn, James B Rawlings, and Moritz Diehl. CasADi – A software framework for nonlinear optimization and optimal control. *Mathematical Programming Computation*, 11(1):1–36, 2019.
3. Johan Andreasson and Tilman Bünte. Global chassis control based on inverse vehicle dynamics models. *Vehicle System Dynamics*, 44(sup1):321–328, 2006.
4. Pierre Apkarian, Pascal Gahinet, and Greg Becker. Self-scheduled $h\infty$ control of linear parameter-varying systems: a design example. *Automatica*, 31(9):1251–1261, 1995.
5. Ienkaran Arasaratnam and Simon Haykin. Cubature Kalman Filters. *IEEE Transactions on Automatic Control*, 54(6):1254–1269, 2009.
6. Mansour Ataei, Amir Khajepour, and Soo Jeon. Reconfigurable integrated stability control for four- and three-wheeled urban vehicles with flexible combinations of actuation systems. *IEEE/ASME Transactions on Mechatronics*, 23(5):2031–2041, 2018.
7. Franffcois Auger, Mickael Hilairet, Josep M Guerrero, Eric Monmasson, Teresa Orlowska-Kowalska, and Seiichiro Katsura. Industrial applications of the Kalman Filter: A review. *IEEE Transactions on Industrial Electronics*, 60(12):5458–5471, 2013.
8. Sofiane Bacha, Ramzi Saadi, Mohamed Yacine Ayad, Abdennacer Aboubou, and Mebarek Bahri. A review on vehicle modeling and control technics used for autonomous vehicle path following. In *2017 International Conference on Green Energy Conversion Systems (GECS)*, pages 1–6, 2017.
9. G Baffet, A Charara, D Lechner, and D Thomas. Experimental evaluation of observers for tire-road forces, sideslip angle and wheel cornering stiffness. *Vehicle System Dynamics*, 46(6):501–520, 2008.
10. Guoxing Bai, Yu Meng, Li Liu, Weidong Luo, Qing Gu, and Li Liu. Review and comparison of path tracking based on model predictive control. *Electronics*, 8(10):1077, 2019.
11. Hanna Bellem, Thorben Schöhenberg, Josef F. Krems, and Michael Schrauf. Objective metrics of comfort: Developing a driving style for highly automated vehicles. *Transportation Research Part F: Traffic Psychology and Behaviour*, 41:45–54, 2016
12. Mohamed Belrzaeg, Abdussalam Ali Ahmed, Amhimmid Q Almabrouk, Mohamed Mohamed Khaleel, Alforjani Ali Ahmed, and Meshaal Almukhtar. Vehicle dynamics and tire models: An overview. *World Journal of Advanced Research and Reviews*, 12(1):331–348, 2021.

13. Beatriz L Boada, María Jesús López Boada, and V Diaz. Vehicle sideslip angle measurement based on sensor data fusion using an integrated ANFIS and an unscented Kalman Filter algorithm. *Mechanical Systems and Signal Processing*, 72:832–845, 2016.
14. Rainer Busch, Klaus Webers, and Achim Seibertz. Ford integrated vehicle dynamics control: concept. *Ford Forschungszentrum Aachen (FFA) Vehicle Electronics & Controls*, 2:26, 2003
15. Yong Xin Cao, Yan Fang Liu, Xiang Yang Xu, and Yang Yang. Simulation for handing and stability of vehicle based on SimulationX. *Advanced Materials Research*, 291:537–541, 2011.
16. Federico Cheli,.Edoardo Sabbioni, M Pesce, and Stefano Melzi. A methodology for vehicle sideslip angle identification: comparison with experimental data. *Vehicle System Dynamics*, 45(6):549–563, 2007.
17. Jianyu Chen, Wei Zhan, and Masayoshi Tomizuka. Autonomous driving motion planning with constrained iterative LQR. *IEEE Transactions on Intelligent Vehicles*, 4(2):244–254, 2019.
18. Jie Chen, Jian Song, Liang Li, Gang Jia, Xu Ran, and Cai Yang. UKF-based adaptive variable structure observer for vehicle sideslip with dynamic correction. *IET Control Theory & Applications*, 10(14):1641–1652, 2016.
19. Liqing Chen, Zhiqiang Li, Juanjuan Yang, and Yu Song. Lateral stability control of four-wheel-drive electric vehicle based on coordinated control of torque distribution and ESP differential braking. *Actuators*, 10(6):135, 2021.
20. Te Chen, Long Chen, Xing Xu, Yingfeng Cai, and Xiaoqiang Sun. Simultaneous path following and lateral stability control of 4WD-4WS autonomous electric vehicles with actuator saturation. *Advances in Engineering Software*, 128:46–54, 2019.
21. Wuwei Chen, Xiutian Liang, Qidong Wang, Linfeng Zhao, and Xiao Wang. Extension coordinated control of four wheel independent drive electric vehicles by AFS and DYC. *Control Engineering Practice*, 101:104504, 2020.
22. Xinbo Chen, Yanqun Han, and Peng Hang. Researches on 4WIS-4WID Stability with LQR Coordinated 4WS and DYC. In Matthijs Klomp, Fredrik Bruzelius, Jens Nielsen, and Angela Hillemyr, editors, *Advances in Dynamics of Vehicles on Roads and Tracks*, pages 1508–1516. Springer International Publishing, Cham, 2020.
23. Shuo Cheng, Liang Li, and Jie Chen. Fusion algorithm design based on adaptive SCKF and integral correction for side-slip angle observation. *IEEE Transactions on Industrial Electronics*, 65(7):5754–5763, 2017.
24. Shuo Cheng, Liang Li, Hong-Qiang Guo, Zhen-Guo Chen, and Peng Song. Longitudinal collision avoidance and lateral stability adaptive control system based on MPC of autonomous vehicles. *IEEE Transactions on Intelligent Transportation Systems*, 21(6):2376–2385, 2020.
25. Shuo Cheng, Liang Li, Hong-Qiang Guo, Zhen-Guo Chen, and Peng Song. Longitudinal collision avoidance and lateral stability adaptive control system based on MPC of autonomous vehicles. *IEEE Transactions on Intelligent Transportation Systems*, 21(6):2376–2385, 2020.
26. Shuo Cheng, Liang Li, Cong-Zhi Liu, Xiuheng Wu, Sheng-Nan Fang, and Jia-Wang Yong. Robust LMI-based H-infinite controller integrating AFS and DYC of autonomous vehicles with parametric uncertainties. *IEEE Transactions on Systems, Man, and Cybernetics: Systems*, 51(11):6901–6910, 2021.

27. Shuo Cheng, Liang Li, Ming-ming Mei, Yu-liang Nie, and Lin Zhao. Multiple-objective adaptive cruise control system integrated with DYC. *IEEE Transactions on Vehicular Technology*, 68(5):4550–4559, 2019.
28. Shuo Cheng, Zheng Wang, Bo Yang, Liang Li, and Kimihiko Nakano. Quantitative evaluation methodology for chassis-domain dynamics performance of automated vehicles. *IEEE Transactions on Cybernetics*, 53(9):5938–5948, 2023.
29. H Cherouat, M Braci, and S Diop. Vehicle velocity, side slip angles and yaw rate estimation. In *Proceedings of the IEEE International Symposium on Industrial Electronics, 2005. ISIE 2005.*, volume 1, pages 349–354. IEEE, 2005.
30. Daniel Chindamo, Basilio Lenzo, and Marco Gadola. On the vehicle sideslip angle estimation: a literature review of methods, models, and innovations. *Applied Sciences*, 8(3):355, 2018.
31. Wanki Cho, Jangyeol Yoon, Seongjin Yim, Bongyeong Koo, and Kyongsu Yi. Estimation of tire forces for application to vehicle stability control. *IEEE Transactions on Vehicular Technology*, 59(2):638–649, 2009.
32. Laurène Claussmann, Marc Revilloud, Dominique Gruyer, and Sébastien Glaser. A review of motion planning for highway autonomous driving. *IEEE Transactions on Intelligent Transportation Systems*, 21(5):1826–1848, 2020.
33. Song Dafeng, Li Jing, Ma Zhimin, Li Youde, Zhao Jian, and Liu Wei. Application of can in vehicle traction control system. In *IEEE International Conference on Vehicular Electronics and Safety*, pages 188–192, 2005.
34. Jamil Dakhlallah, SŽebastien Glaser, Said Mammar, and Yazid Sebsadji. Tireroad forces estimation using extended Kalman filter and sideslip angle evaluation. In *2008 American Control Conference*, pages 4597–4602. IEEE, 2008.
35. Zejian Deng, Duanfeng Chu, Chaozhong Wu, Shidong Liu, Chen Sun, Teng Liu, and Dongpu Cao. A probabilistic model for driving-style-recognition-enabled driver steering behaviors. *IEEE Transactions on Systems, Man, and Cybernetics: Systems*, 52(3):1838–1851, 2022.
36. Anup Doshi and Mohan M. Trivedi. Examining the impact of driving style on the predictability and responsiveness of the driver: Real-world and simulator analysis. In *2010 IEEE Intelligent Vehicles Symposium*, pages 232–237, June 2010.
37. Moustapha Doumiati, Alessandro Correa Victorino, Ali Charara, and Daniel Lechner. Onboard real-time estimation of vehicle lateral tire-road forces and sideslip angle. *IEEE/ASME Transactions on Mechatronics*, 16(4):601–614, 2010.
38. Haiping Du, James Lam, Kie-Chung Cheung, Weihua Li, and Nong Zhang. Side-slip angle estimation and stability control for a vehicle with a non-linear tyre model and a varying speed. *Proceedings of the Institution of Mechanical Engineers, Part D: Journal of Automobile Engineering*, 229(4):486–505, 2015.
39. Xianping Du, Chao Sun, Yiang Zheng, Xiexing Feng, and Na Li. Evaluation of vehicle vibration comfort using deep learning. *Measurement*, 173:108634, 2021.
40. Howard Dugoff, Paul S Fancher, and Leonard Segel. An analysis of tire traction properties and their influence on vehicle dynamic performance. *SAE transactions*, pages 1219–1243, 1970.

41. Mohamed Elbanhawi and Milan Simic. Sampling-based robot motion planning: A review. *IEEE Access*, 2:56–77, 2014.
42. Samira Eshghi and Renuganth Varatharajoo. Nonsingular terminal sliding mode control technique for attitude tracking problem of a small satellite with combined energy and attitude control system (CEACS). *Aerospace Science and Technology*, 76:14–26, 2018.
43. Charles M Farmer. Effect of electronic stability control on automobile crash risk. *Traffic Injury Prevention*, 5(4):317–325, 2004.
44. Dave Ferguson, Thomas M Howard, and Maxim Likhachev. Motion planning in urban environments. *Journal of Field Robotics*, 25(11-12):939–960, 2008.
45. Yoshimi Furukawa and Masato Abe. Advanced chassis control systems for vehicle handling and active safety. *Vehicle System Dynamics*, 28(2-3):59–86, 1997.
46. David González, Joshué Pérez, Vicente Milanés, and Fawzi Nashashibi. A review of motion planning techniques for automated vehicles. *IEEE Transactions on Intelligent Transportation Systems*, 17(4):1135–1145, 2015.
47. David González, Joshué Pérez, Vicente Milanés, and Fawzi Nashashibi. A review of motion planning techniques for automated vehicles. *IEEE Transactions on Intelligent Transportation Systems*, 17(4):1135–1145, 2016.
48. Robert Grepl, Josef Vejlupek, Vojtech Lambersky, Michal Jasansky, Filip Vadlejch, and Pavel Coupek. Development of 4WS/4WD experimental vehicle: Platform for research and education in mechatronics. In *2011 IEEE International Conference on Mechatronics*, pages 893–898, Istanbul, Turkey, April 2011. IEEE.
49. Håvard Fjær Grip, Lars Imsland, Tor A Johansen, Thor I Fossen, Jens C Kalkkuhl, and Avshalom Suissa. Nonlinear vehicle side-slip estimation with friction adaptation. *Automatica*, 44(3):611–622, 2008.
50. Kang Gui, Lin Ye, Junfeng Ge, Faouzi Alaya Cheikh, and Lizhen Huang. Road surface condition detection utilizing resonance frequency and optical technologies. *Sensors and Actuators A: Physical*, 297:111540, 2019.
51. Hongyan Guo, Zhenyu Yin, Dongpu Cao, Hong Chen, and Chen Lv. A review of estimation for vehicle tire-road interactions toward automated driving. *IEEE Transactions on Systems, Man, and Cybernetics: Systems*, 49(1):14–30, 2019.
52. Huihui Guo, Fan Wu, Yunchuan Qin, Ruihui Li, Keqin Li, and Kenli Li. Recent trends in task and motion planning for robotics: A survey. *ACM Computing Surveys*, 55(13s):1–36, 2023.
53. Konghui Guo and Lei Ren. A unified semi-empirical tire model with higher accuracy and less parameters. *SAE Transactions*, pages 1513–1520, 1999.
54. Fredrik Gustafsson. Slip-based tire-road friction estimation. *Automatica*, 33(6):1087–1099, 1997.
55. Kyoungseok Han, Yoonjin Hwang, Eunjae Lee, and S Choi. Robust estimation of maximum tire-road friction coefficient considering road surface irregularity. *International Journal of Automotive Technology*, 17:415–425, 2016.
56. Peng Hang, Chen Lv, Chao Huang, Jiacheng Cai, Zhongxu Hu, and Yang Xing. An iIntegrated framework of decision making and motion planning for autonomous vehicles considering social behaviors. *IEEE Transactions on Vehicular Technology*, 69(12):14458–14469, 2020.

57. Peng Hang, Xin Xia, Guang Chen, and Xinbo Chen. Active Safety Control of Automated Electric Vehicles at Driving Limits: A Tube-based MPC Approach. *IEEE Transactions on Transportation Electrification*, 8(1): 1338–1349, 2021.
58. Peter E Hart, Nils J Nilsson, and Bertram Raphael. A formal basis for the heuristic determination of minimum cost paths. *IEEE Transactions on Systems Science and Cybernetics*, 4(2):100–107, 1968.
59. Tesheng Hsiao. Robust estimation and control of tire traction forces. *IEEE Transactions on Vehicular Technology*, 62(3):1378–1383, 2012.
60. Xuemin Hu, Long Chen, Bo Tang, Dongpu Cao, and Haibo He. Dynamic path planning for autonomous driving on various roads with avoidance of static and moving obstacles. *Mechanical Systems and Signal Processing*, 100:482–500, 2018.
61. Chao Huang, Hailong Huang, Peng Hang, Hongbo Gao, Jingda Wu, Zhiyu Huang, and Chen Lv. Personalized trajectory planning and control of lane-change maneuvers for autonomous driving. *IEEE Transactions on Vehicular Technology*, 70(6):5511–5523, 2021.
62. Xiaoyu Huang and Junmin Wang. Center of gravity height real-time estimation for lightweight vehicles using tire instant effective radius. *Control Engineering Practice*, 21(4):370–380, 2013.
63. Yiwen Huang and Yan Chen. Vehicle lateral stability control based on shiftable stability regions and dynamic margins. *IEEE Transactions on Vehicular Technology*, 69(12):14727–14738, 2020.
64. Joo Young Hwang, Jun Song Kim, Sang Seok Lim, and Kyu Ho Park. A fast path planning by path graph optimization. *IEEE Transactions on Systems, Man, and Cybernetics-Part A: Systems and Humans*, 33(1):121–129, 2003.
65. Hocine Imine, Leonid Fridman, Hassan Shraim, and Mohamed Djemai. *Sliding Mode Based Analysis and Identification of Vehicle Dynamics*, volume 414. Springer Science & Business Media, 2011.
66. Shoji Inagaki, Ikuo Kushiro, and Masaki Yamamoto. Analysis on vehicle stability in critical cornering using phase-plane method. *JSAE Review*, 2(16):216,1995.
67. K Indu and M Aswatha Kumar. Electric vehicle control and driving safety systems: A review. *IETE Journal of Research*, 69(1):482–498, 2023.
68. K. Ito and K. Xiong. Gaussian filters for nonlinear filtering problems. *IEEE Transactions on Automatic Control*, 45(5):910–927, 2000.
69. Dhruv Jagga, Maolong Lv, and Simone Baldi. Hybrid adaptive chassis control for vehicle lateral stability in the presence of uncertainty. In *2018 26th Mediterranean Conference on Control and Automation (MED)*, pages 1–6, Zadar, Croatia, June 2018. IEEE.
70. Reza N. Jazar. *Vehicle Planar Dynamics: Theory and Applications*. Springer, New York, NY, corrected at 3. printing edition, 2009.
71. Guirong Jiang, Lifu Liu, Changhong Guo, Jie Chen, Fahad Muhammad, and Xiaolu Miao. A novel fusion algorithm for estimation of the side-slip angle and the roll angle of a vehicle with optimized key parameters. *Proceedings of the Institution of Mechanical Engineers, Part D: Journal of Automobile Engineering*, 231(2):161–174, 2017.

72. Xinwei Jiang, Long Chen, Xing Xu, Yingfeng Cai, Yong Li, and Wujie Wang. Analysis and optimization of energy efficiency for an electric vehicle with four independent drive in-wheel motors. *Advances in Mechanical Engineering*, 10(3):168781401876554, 2018.
73. Xianjian Jin and Guodong Yin. Estimation of lateral tire-road forces and sideslip angle for electric vehicles using interacting multiple model filter approach. *Journal of the Franklin Institute*, 352(2):686–707, 2015.
74. Hui Jing, Rongrong Wang, Junmin Wang, and Nan Chen. Robust H∞ dynamic output-feedback control for four-wheel independently actuated electric ground vehicles through integrated AFS/DYC. *Journal of the Franklin Institute*, 355(18):9321–9350, 2018.
75. Zhiyang Ju, Hui Zhang, Xiang Li, Xiaoguang Chen, Jinpeng Han, and Manzhi Yang. A survey on attack detection and resilience for connected and automated vehicles: From vehicle dynamics and control perspective. *IEEE Transactions on Intelligent Vehicles*, 7(4):815–837, 2022.
76. Mustafa Ali Arat Kanwar Bharat Singh and Saied Taheri. Literature review and fundamental approaches for vehicle and tire state estimation. *Vehicle System Dynamics*, 57(11):1643–1665, 2019.
77. Larissa Labakhua, Urbano Nunes, Rui Rodrigues, and Fátima S Leite. Smooth trajectory planning for fully automated passengers vehicles: Spline and clothoid based methods and its simulation. In *Informatics in Control Automation and Robotics: Selected Papers from the International Conference on Informatics in Control Automation and Robotics 2006*, pages 169–182. Springer, 2008.
78. Chankyu Lee, Karl Hedrick, and Kyongsu Yi. Real-time slip-based estimation of maximum tire-road friction coefficient. *IEEE/ASME Transactions on Mechatronics*, 9(2):454–458, 2004.
79. Arie Levant. Homogeneity approach to high-order sliding mode design. *Automatica*, 41(5):823–830, 2005.
80. Arie Levant. Principles of 2-sliding mode design. *Automatica*, 43(4):576–586, 2007.
81. Junqing Li, Xiangyang Xu, Wantong Zhao, Peishen Zhao, Hanning Zhang, Shuhan Wang, and Peng Dong. Longitudinal tire force estimation for a novel corner drive system based on $\ell\infty$ robust observer. In *2024 IEEE 22nd International Conference on Industrial Informatics (INDIN)*, pages 1–4. IEEE, 2024.
82. Liang Li, Gang Jia, Jie Chen, Hongjun Zhu, Dongpu Cao, and Jian Song. A novel vehicle dynamics stability control algorithm based on the hierarchical strategy with constrain of nonlinear tyre forces. *Vehicle System Dynamics*, 53(8):1093–1116, 2015.
83. Liang Li, Gang Jia, Xu Ran, Jian Song, and Kaihui Wu. A variable structure extended Kalman filter for vehicle sideslip angle estimation on a low friction road. *Vehicle System Dynamics*, 52(2):280–308, 2014.
84. Xiaohui Li, Zhenping Sun, Dongpu Cao, Zhen He, and Qi Zhu. Real-time trajectory planning for autonomous urban driving: Framework, algorithms, and verifications. *IEEE/ASME Transactions on Mechatronics*, 21(2):740–753, 2016.

85. Xiaohui Li, Zhenping Sun, Dongpu Cao, Daxue Liu, and Hangen He. Development of a new integrated local trajectory planning and tracking control framework for autonomous ground vehicles. *Signal Processing and Control challenges for Smart Vehicles*, 87:118–137, 2017.
86. Xu Li, Ching-Yao Chan, and Yu Wang. A reliable fusion methodology for simultaneous estimation of vehicle sideslip and yaw angles. *IEEE Transactions on Vehicular Technology*, 65(6):4440–4458, 2015.
87. Ying Li, Lingfei Ma, Zilong Zhong, Fei Liu, Michael A. Chapman, Dongpu Cao, and Jonathan Li. Deep learning for lidar point clouds in autonomous driving: A review. *IEEE Transactions on Neural Networks and Learning Systems*, 32(8):3412–3432, 2021 .
88. Wonteak Lim, Seongjin Lee, Myoungho Sunwoo, and Kichun Jo. Hybrid trajectory planning for autonomous driving in on-road dynamic scenarios. *IEEE Transactions on Intelligent Transportation Systems*, 22(1):341–355, 2021
89. Jia Liu, Zhiheng Yang, Zhejun Huang, Wenfei Li, Shaobo Dang, and Huiyun Li. Simulation performance evaluation of pure pursuit, Stanley, LQR, MPC controller for autonomous vehicles. In *2021 IEEE International Conference on Real-Time Computing and Robotics (RCAR)*, pages 1444–1449, Xining, China, July 2021. IEEE.
90. Kang-Zhi Liu. LMI approach to robust control. In *Encyclopedia of Systems and Control*, pages 1152–1158. Springer, 2021.
91. Y-H Liu, T Li, Y-Y Yang, X-W Ji, and J Wu. Estimation of tire-road friction coefficient based on combined APF-IEKF and iteration algorithm. *Mechanical Systems and Signal Processing*, 88:25–35, 2017.
92. Chen Lv, Xiaosong Hu, Alberto Sangiovanni-Vincentelli, Yutong Li, Clara Marina Martinez, and Dongpu Cao. Driving-style-based codesign optimization of an automated electric vehicle: A cyber-physical system approach. *IEEE Transactions on Industrial Electronics*, 66(4):2965–2975, 2019.
93. Victor Mazzilli, Stefano De Pinto, Leonardo Pascali, Michele Contrino, Francesco Bottiglione, Giacomo Mantriota, Patrick Gruber, and Aldo Sorniotti. Integrated chassis control: Classification, analysis and future trends. *Annual Reviews in Control*, 51:172–205, 2021.
94. Sky McKinley and Megan Levine. Cubic spline interpolation. *College of the Redwoods*, 45(1):1049–1060, 1998.
95. Rakuzou Mitamura, Masanori Tani, Tadao Tanaka, and Miroo Yuasa. System integration for new mobility. *SAE Transactions*, pages 1418–1428, 1988.
96. MG Mohanan and Ambuja Salgoankar. A survey of robotic motion planning in dynamic environments. *Robotics and Autonomous Systems*, 100:171–185, 2018.
97. Mette Møller and Sonja Haustein. Keep on cruising: Changes in lifestyle and driving style among male drivers between the age of 18 and 23. *Transportation Research Part F: Traffic Psychology and Behaviour*, 20:59–69, 2013.
98. Jaime A Moreno and Marisol Osorio. Strict Lyapunov functions for the super-twisting algorithm. *IEEE Transactions on Automatic Control*, 57(4):1035–1040, 2012.
99. Masao Nagai, Yutaka Hirano, and Sachiko Yamanaka. Integrated control of active rear wheel steering and direct yaw moment control. *Vehicle System Dynamics*, 27(5-6):357–370, 1997.

100. Jun Ni, Jibin Hu, and Changle Xiang. Envelope control for four-wheel independently actuated autonomous ground vehicle through AFS/DYC integrated control. *IEEE Transactions on Vehicular Technology*, 66(11):9712–9726, 2017.
101. Jaswanth Nidamanuri, Chinmayi Nibhanupudi, Rolf Assfalg, and Hrishikesh Venkataraman. A progressive review: Emerging technologies for ADAS driven solutions. *IEEE Transactions on Intelligent Vehicles*, 7(2):326–341, 2022.
102. Donghong Ning, Haiping Du, Nong Zhang, Shuaishuai Sun, and Weihua Li. Controllable electrically interconnected suspension system for improving vehicle vibration performance. *IEEE/ASME Transactions on Mechatronics*, 25(2):859–871, 2020.
103. Hiroki Ohara and Toshiyuki Murakami. A stability control by active angle control of front-wheel in a vehicle system. *IEEE Transactions on Industrial Electronics*, 55(3):1277–1285, 2008.
104. H. Pacejka. *Tyre and Vehicle Dynamics*. Society of Automotive Engineers, Inc, Warrendale, PA, 2002.
105. Hans Pacejka. *Tire and Vehicle Dynamics*. Elsevier, 2005.
106. Hans B Pacejka and Egbert Bakker. The magic formula tyre model. *Vehicle System Dynamics*, 21(S1):1–18, 1992.
107. Giseo Park, Kyoungseok Han, Kanghyun Nam, Heeseong Kim, and Seibum B. Choi. Torque vectoring algorithm of electronic-four-wheel drive vehicles for enhancement of cornering performance. *IEEE Transactions on Vehicular Technology*, 69(4):3668–3679, 2020.
108. Haonan Peng, Weida Wang, Quan An, Changle Xiang, and Liang Li. Path tracking and direct yaw moment coordinated control based on robust MPC with the finite time horizon for autonomous independent-drive vehicles. *IEEE Transactions on Vehicular Technology*, 69(6):6053–6066, 2020.
109. Wilfrid Perruquetti and Jean Pierre Barbot. *Sliding Mode Control in Engineering*, volume 11. Marcel Dekker New York, 2002.
110. Aurelio Piazzi, CG Lo Bianco, Massimo Bertozzi, Alessandra Fascioli, and Alberto Broggi. Quintic G/sup 2/-splines for the iterative steering of vision-based autonomous vehicles. *IEEE Transactions on Intelligent Transportation Systems*, 3(1):27–36, 2002.
111. Gengxin Qi, Xiaobin Fan, and Hao Li. A review on identification methods of road friction coefficient. *Recent Patents on Engineering*, 16(2):75–90, 2022.
112. Yadollah Rasekhipour, Amir Khajepour, Shih-Ken Chen, and Bakhtiar Litkouhi. A potential field-based model predictive path-planning controller for autonomous road vehicles. *IEEE Transactions on Intelligent Transportation Systems*, 18(5):1255–1267, 2017.
113. Joshue Perez Rastelli, Ray Lattarulo, and Fawzi Nashashibi. Dynamic trajectory generation using continuous-curvature algorithms for door to door assistance vehicles. In *2014 IEEE Intelligent Vehicles Symposium Proceedings*, pages 510–515. IEEE, 2014.
114. Laura Ryan Ray. Nonlinear state and tire force estimation for advanced vehicle control. *IEEE Transactions on Control Systems Technology*, 3(1):117–124, 1995.
115. B Dayanand Reddy. *Introductory Functional Analysis: With Applications to Boundary Value Problems and Finite Elements*, volume 27. Springer Science & Business Media, 2013.

116. Ayyoub Rezaeian, Reza Zarringhalam, Saber Fallah, Wael Melek, Amir Khajepour, S-Ken Chen, N Moshchuck, and Bakhtiar Litkouhi. Novel tire force estimation strategy for real-time implementation on vehicle applications. *IEEE Transactions on Vehicular Technology*, 64(6):2231–2241, 2014.
117. Christina Ršodel, Susanne Stadler, Alexander Meschtscherjakov, and Manfred Tscheligi. Towards autonomous cars: The effect of autonomy levels on acceptance and user experience. In *Proceedings of the 6th International Conference on Automotive User Interfaces and Interactive Vehicular Applications*, AutomotiveUI'14, pages 1–8, New York, NY, USA, 2014. Association for Computing Machinery.
118. I. Salgado, S. Kamal, I. Chairez, B. Bandyopadhyay, and L. Fridman. Super-twisting-like algorithm in discrete time nonlinear systems. In *International Conference on Advanced Mechatronic Systems*, Zhengzhou, China, 2011. IEEE.
119. Yukihisa Sasada, M Nakano, Y Miyata, K Sugiyama, and Eiichi Shiba. Development of the road surface condition sensing system. In *Proceedings 199 IEEE/IEEJ/JSAI International Conference on Intelligent Transportation Systems (Cat. No. 99TH8383)*, pages 14–19. IEEE, 1999.
120. RS Sharp and Huei Peng. Vehicle dynamics applications of optimal control theory. *Vehicle System Dynamics*, 49(7):1073–1111, 2011.
121. Shuiwen Shen, Jun Wang, Peng Shi, and Giuliano Premier. Nonlinear dynamics and stability analysis of vehicle plane motions. *Vehicle System Dynamics*, 45(1):15–35, 2007.
122. Yunyi Shen, Jian Liu, and Yasong Luo. Review of path planning algorithms for unmanned vehicles. In *2021 IEEE 2nd International Conference on Information Technology, Big Data and Artificial Intelligence (ICIBA)*, volume 2, pages 400–405, 2021.
123. Yuri Shtessel, Christopher Edwards, Leonid Fridman, Arie Levant, et al. *Sliding Mode Control and Observation*, volume 10. Springer, 2014.
124. Sebastien Solvar, Vu Le, Malek Ghanes, J. P. Barbot, and Gaëtan Santomenna. Sensorless second order sliding mode observer for induction motor. In *IEEE International Conference on Control Applications*, pages 1933–1938, Yokohama, Japan, 2010. IEEE.
125. Subaselvi Sundarraj, R Vijaya Kumar Reddy, B Mahesh Babu, Gururaj Harinahalli Lokesh, Francesco Flammini, and Rajesh Natarajan. Route planning for an autonomous robotic vehicle employing a weight-controlled particle swarm-optimized Dijkstra algorithm. *IEEE Access*, 11:92433–92442, 2023.
126. R Talebitooti and Mansour Torabi. Identification of tire force characteristics using a hybrid method. *Applied Soft Computing*, 40:70–85, 2016.
127. H Tanaka, H Inoue, and H Iwata. Development of a vehicle integrated control system. In *XXIV FISITA Congress*, page 925049, London, United Kingdom, 1992. The Automotive Technology Serving Society.
128. A. T. van Zanten. Bosch ESP systems: 5 years of experience. *SAE Transactions*, 109:428–436, 2000.
129. A. T. Van Zanten. Bosch ESP Systems: 5 years of experience. In *SAE 2000 Automotive Dynamics & Stability Conference*, pages 1–9, May 2000.
130. Anton T Van Zanten. Bosch ESP systems: 5 years of experience. *SAE Transactions*, pages 428–436, 2000.

131. Anshu Vashisth, Ranbir Singh Batth, and Rupert Ward. Existing path planning techniques in unmanned aerial vehicles (UAVs): A systematic review. In *2021 International Conference on Computational Intelligence and Knowledge Economy (ICCIKE)*, pages 366–372, March 2021.
132. Richard Vinter. *Optimal Control*. Birkhäuser Boston, Boston, 2010.
133. Junmin Wang and Raul G Longoria. Coordinated and reconfigurable vehicle dynamics control. *IEEE Transactions on Control Systems Technology*, 17(3):723–732, 2009.
134. Junnian Wang, Shoulin Gao, Kai Wang, Yan Wang, and Qingsong Wang. Wheel torque distribution optimization of four-wheel independent-drive electric vehicle for energy efficient driving. *Control Engineering Practice*, 110:104779, 2021.
135. Guo Wei, Xu Xiangyang, Cao Yongxin, and Yang Yang. Simulation of powertrain and dynamics of automobile based on simulationx. In *2011 6th IEEE Conference on Industrial Electronics and Applications*, pages 2326–2330. IEEE, 2011.
136. Zhang Wei and Guo Xuexun. An abs control strategy for commercial vehicle. *IEEE/ASME Transactions on Mechatronics*, 20(1):384–392, 2015.
137. Wang Weida, Ding Nenggen, and Xu Xiangyang. An improved self-adaptive algorithm of vehicle reference speeds for ABS. In *2006 IEEE International Conference on Vehicular Electronics and Safety*, pages 98–102. IEEE, 2006.
138. Wang Weida, Ding Nenggen, Xu Xiangyang, and Zhang Jian. Simulation of the control strategy of vehicular ASR system under the circumstance of simulink. In *2007 IEEE International Conference on Automation and Logistics*, pages 2749–2753. IEEE, 2007.
139. Moritz Werling, Julius Ziegler, Sšoren Kammel, and Sebastian Thrun. Optimal trajectory generation for dynamic street scenarios in a Frenet Frame. In *2010 IEEE International Conference on Robotics and Automation*, pages 987–993, Anchorage, AK, May 2010. IEEE
140. MA Wilkin, WJ Manning, DA Crolla, and MC Levesley. Use of an extended Kalman filter as a robust tyre force estimator. *Vehicle System Dynamics*, 44(sup1):50–59, 2006.
141. Wenda Xu, Junqing Wei, John M. Dolan, Huijing Zhao, and Hongbin Zha. A real-time motion planner with trajectory optimization for autonomous vehicles. In *2012 IEEE International Conference on Robotics and Automation*, pages 2061–2067, 2012.
142. Xiangyang Xu, Dawei Zhang, Junqing Li, Hangning Zhang, Peishen Zhao, Zhuochao Liu, Peng Dong, Yanfang Liu, and Shuhan Wang. Design and control of semi-active suspension for a novel corner drive system. In *2024 8th CAA International Conference on Vehicular Control and Intelligence (CVCI)*, pages 1–6. IEEE, 2024.
143. Shuai Yan, Weichao Sun, Xinghu Yu, and Huijun Gao. Adaptive sensor fault accommodation for vehicle active suspensions via partial measurement information. *IEEE Transactions on Cybernetics*, 52(11):12290–12301, 2022.
144. Yu-guang Yan, Hong-guo Xu, and Hong-fei Liu. Estimating vehicle stability region based on energy function. *Discrete Dynamics in Nature and Society*, 2015:805063, 2015.
145. Shaopu Yang, Yongjie Lu, and Shaohua Li. An overview on vehicle dynamics. *International Journal of Dynamics and Control*, 1:385–395, 2013.

146. Guo-Dong Yin. Estimating lateral stability region for four wheel independently – Actuated electric vehicle considering steering. In *SAE 2013 Commercial Vehicle Engineering Congress*. SAE International, September 2013.
147. Seung-Han You, Jin-Oh Hahn, and Hyeongcheol Lee. New adaptive approaches to real-time estimation of vehicle sideslip angle. *Control Engineering Practice*, 17(12):1367–1379, 2009.
148. Fan Yu, Dao-Fei Li, and DA Crolla. Integrated vehicle dynamics control—state-of-the art review. In *2008 IEEE Vehicle Power and Propulsion Conference*, pages 1–6. IEEE, 2008.
149. Han-ye Zhang, Wei-ming Lin, and Ai-xia Chen. Path planning for the mobile robot: A review. *Symmetry*, 10(10), 2018.
150. Hui Zhang and Junmin Wang. Vehicle lateral dynamics control through AFS/DYC and robust Gain-Scheduling approach. *IEEE Transactions on Vehicular Technology*, 65(1):489–494, 2016.
151. Jiaxu Zhang. *Research on Integrated Nonlinear Robust Control Method for Automotive Chassis (In Chinese)*. PhD thesis, Jilin University, 2018.
152. Jing Zhao, Pak Kin Wong, Xinbo Ma, and Zhengchao Xie. Chassis integrated control for active suspension, active front steering and direct yaw moment systems using hierarchical strategy. *Vehicle System Dynamics*, 55(1):72–103, 2017.
153. Lihang Zhao, Jin Huang, He Liu, Bingnan Li, and Wubin Kong. Second-order sliding-mode observer with online parameter identification for sensorless induction motor drives. *IEEE Transactions on Industrial Electronics*, 61(10):5280–5289, 2014.
154. WanZhong Zhao, MiLi Fan, ChunYan Wang, Zhilin Jin, and Yufang Li. H∞/extension stability control of automotive active front steering system. *Mechanical Systems and Signal Processing*, 115:621–636, 2019.
155. Min Zhu, Huiyan Chen, and Guangming Xiong. A model predictive speed tracking control approach for autonomous ground vehicles. *Mechanical Systems and Signal Processing*, 87:138–152, 2017.
156. YU Zhuoping, FENG Yuan, Xiong Lu, et al. Review on vehicle dynamics control of distributed drive electric vehicle. *Journal of Mechanical Engineering*, 49(8):105–114, 2013.

Index

Made in the USA
Columbia, SC
17 September 2026

c12bba8e-1242-43f9-9c70-a0a9ae9a3ab7R01